Lecture Notes in Mathematics

An informal series of special lectures, seminars and reports on mathematical topics

Edited by A. Dold, Heidelberg and B. Eckmann, Zürich

23

P. L. Ivănescu · S. Rudeanu

Institute of Mathematics

Academy of S.R. Romania, Bucharest

Pseudo-Boolean Methods for Bivalent Programming

Lecture at the First European Meeting of the Institute of Management Sciences and of the Econometric Institute, Warsaw, September 2 – 7, 1966

1966

Springer-Verlag · Berlin · Heidelberg · New York

Library of Congress Catalog Card Number 66-27979. Title No. 7343.

TABLE OF CONTENTS

PREFACE

The aim of the present lecture is to propose a method for bivalent (0,1) linear and nonlinear programming.

We define a pseudo-Boolean function as a real-valued function with bivalent arguments. An equation (inequality) whose members are pseudo-Boolean functions is named pseudo-Boolean. In this lecture procedures will be proposed for : 1) solving systems of linear pseudo-Boolean equations and inequalities; 2)solving systems of nonlinear pseudo-Boolean equations and inequalities; 3) minimizing a pseudo - Boolean function with or without constraints.

The material of this lecture will form the core of a book on "Boolean Methods in Operations Research and Related Areas",to appear in the "Econometrics and Operations Research" series of the Springer - Verlag.

PART I.

LINEAR PSEUDO-BOOLEAN EQUATIONS AND INEQUALITIES

Numerous problems in operations research may be regarded as programming problems with bivalent variables. Typical cases leading to this model were described in G.B. DANTZIG's paper /10/, /11/ (see also /12/) on the importante of integer programming.

In 1958, R.E.GOMORY /18/ (see also /19/) has given a method for solving integer linear programs; since that time, this field has known a constant development.

Of course, the general methods for solving integer linear programs are applicable to zero-one problems; however, special methods using the particularities of this cases were loocked for.

A subclass of the latter methods is based on Boolean techniques and it was R.FORTET /15/,/16/,/17/ who first pointed out this type of approach. Meanwhile the field of problems which can be solved by Boolean methods was enlarged by P.CAMION /6/,/7/, M.CARVALLO /8/,/9/, M.DENIS-PAPIN, R.FAURE and A.KAUFMANN /13/, R.FAURE and Y.MALGRANGE /14/, A.KAUFMANN /27/, A. KAUFMANN and Y.MALGRANGE /28/, K.MAGHOUT /29/, B.ROY /30/, and others. Another type of approach is that of dynamic programming, which was shown by R.BELLMAN /2/ to be applicable for solving certain combinatorial problems.

In 1963, I.ROSENBERG and the present authors, using the basic idea of R.FORTET /16/, have suggested a Boolean method for finding the minima of an integer - (or real-) valued function with bivalent (0 or 1) variables ("pseudo-Boolean function") the variables being possibly subject to certain constraints /25/, /26/ (see also /23/). As a matter of fact, the proposed method was later shown /24/ to be a combination of the dynamic programming approach with Boolean techniques. This method which we call "pseudo-Boolean programming", was then applied for solving numerous problems: integer polynomial programming, problems of graph theory, transportation problems, flows in networks, minimal decompositions of partially ordered sets into chains, etc. (see the expository paper /22/, in which the various applications of the method found up to 1965 are described).

The first part of this lecture starting with some basic definitions and properties of pseudo-Boolean functions, describes a procedure for solving (systems of) linear pseudo-Boolean equations and inequalities (i.e. equations (inequalities) whose sides are pseudo-Boolean functions). This research has an obvious intrinsic interest (see, for instance, the papers of R. FORTET /15/, /16/, /17/, P.CAMION /6/, /7/ and P.L.IVANESCU /20/, /21/) but it is mainly aimed to serve as a tool in the subsequent parts.

The next part will deal with the case of (systems of) nonlinear pseudo-Boolean equations and inequalities.

In the third part, we shall give a procedure for finding the minimum value of a pseudo-Boolean function and all the minimizing points.

In the first part, the approach is mainly combinatorial, while in the following ones we shall lay a stress on Boolean techniques.

A. PROPERTIES OF PSEUDO-BOOLEAN FUNCTIONS

Let B_2 be the two-element Boolean algebra, that is the set $\{0,1\}$, together with the following three operations: the disjunction ($\cup$), defined by

(1)

$\cup$	0	1
0	0	1
1	1	1

,

the conjunction ($\cdot$ or juxtaposition), defined by

(2)

$\cdot$	0	1
0	0	0
1	0	1

and the negation ($^{-}$), defined by

(3)

x	0	1
$\bar{x}$	1	0

§ 1. Pseudo-Boolean Functions

Definition 1. Let R be the field of reals; by a *pseudo-Boolean function*, we shall mean a function

(4) $$f : B_2^n \longrightarrow R ,$$

where B_2^n denotes the cartesian product $\underbrace{B_2 \times \ldots \times B_2}_{n \text{ times}}$.

In other words, a pseudo-Boolean function is simply a real-valued*) function of bivalent (0,1) variables.

R.FORTET calls these functions "integer algebraic functions". Our term is justified by the following remark : if the elements 0 and 1 of B_2 are identified with the reals 0 and 1 - and this will be tacitly assumed in the sequel - then every Boolean function

$$(5) \qquad \varphi : B_2^n \longrightarrow B_2$$

is also a pseudo-Boolean function.

The above remark about the embedding of the set B_2 into the set of integers may be sharpened as follows: the conjunction (2) coincides with the ordinary multiplication between the numbers 0 and 1, while

$$(6) \qquad x \cup y = x + y - xy$$

$$(7) \qquad \bar{x} = 1 - x$$

(the proof reduces to the verification of these equalities for all possible values given to x and y).

It follows that every Boolean expression may be written in terms of the arithmetical operations (by repeated applications of the above rules).

As concerns the pseudo-Boolean functions, let us notice first that such a function is always linear in each of its variables. Indeed, if we set

$$(8) \quad g(x_1,\dots,x_{i-1},x_{i+1},\dots,x_n) = f(x_1,\dots,x_{i-1},1,x_{i+1},\dots,x_n) - f(x_1,\dots,x_{i-1},0,x_{i+1},\dots,x_n) ,$$

*) As a matter of fact, most of the problems occurring in practice involve pseudo-Boolean functions with <u>integer</u> values. Therefore the examples will be of this type.

(9) $h(x_1,\ldots,x_{i-1},x_{i+1},x,\ldots,x_n)=f(x_1,\ldots,x_{i-1},0,x_{i+1},\ldots,x_n)$,

then

(10) $$f(x_1,\ldots,x_n) = x_i g(x_1,\ldots,x_{i-1},x_{i+1},\ldots,x_n) + h(x_1,\ldots,x_{i-1},x_{i+1},\ldots,x_n);$$

conversely, relation (10) implies (9) and (8).

More generally we have the following result, due to T.GASPAR:

THEOREM 1. _Every pseudo-Boolean function may be written as a polynomial, linear in each variable and which, after the reduction of the similar terms, is uniquely determined up to the order of the sums and products._

The _proof_ by induction is immediate.

On the other hand, every pseudo-Boolean function has also a development, analogous to the canonical disjunctive form of a Boolean function.

Setting

(11) $$x^1 = x, \quad x^0 = \bar{x},$$

we have the following:

THEOREM 2.*) _Every pseudo-Boolean function may be written in the form_

(12) $$f(x_1,\ldots,x_n) = \sum_{\alpha_1,\ldots,\alpha_n} c_{\alpha_1\ldots\alpha_n} x_1^{\alpha_1} \ldots x_n^{\alpha_n},$$

where the sum $\sum_{\alpha_1,\ldots,\alpha_n}$ _is extended over the 2^n values of the_

*) Also stated by M. CARVALLO (191, p.lo2) in a matrix formulation.

<u>vector</u> $(\alpha_1,\ldots,\alpha_n)\in B_2^n$, <u>the coefficients</u> $c_{\alpha_1\ldots\alpha_n}$ <u>being uniquely determined by the relations</u>

(13) $$c_{\alpha_1\ldots\alpha_n} = f(\alpha_1,\ldots,\alpha_n).$$

The <u>proof</u> is analogous to the demonstration of the corresponding theorem for Boolean functions.

<u>Example 1.</u> The pseudo-Boolean function

(14) $$f(x_1,x_2,x_3) = 2x_1\bar{x}_2 + 6x_1x_3 - 5\bar{x}_2\bar{x}_3,$$

which can be also defined by the following

- Table 1 -

x_1	x_2	x_3	$f(x_1,x_2,x_3)$
0	0	0	-5
0	0	1	0
0	1	0	0
0	1	1	0
1	0	0	-3
1	0	1	8
1	1	0	0
1	1	1	6

may be also written in the form

(14') $$f(x_1,x_2,x_3) = -5 + 2x_1 + 5x_2 + 5x_3 - 2x_1x_2 + 6x_1x_3 - 5x_2x_3,$$

which is obtained from (14) via formula (7); the function f can be also written in the form

(14") $$f(x_1,x_2,x_3) = -5\bar{x}_1\bar{x}_2\bar{x}_3 - 3x_1\bar{x}_2\bar{x}_3 + 8x_1\bar{x}_2x_3 + 6x_1x_2x_3,$$

which is obtained from the above table by applying Theorem 2.

* * *

B. PSEUDO-BOOLEAN EQUATIONS AND INEQUALITIES

Pseudo-Boolean equations, that is equations of the form

(15) $$f(x_1,\ldots,x_n) = 0,$$

where f are pseudo-Boolean functions, have been studied by R. FORTET /15/,/16/,/17/, P.CAMION /6/,/7/, and P.L.IVANESCU /20/, /21/. These authors have suggested various methods which express the general solution in a parametric form.

The procedure we shall suggest below will offer the sought solutions either completely listed, or grouped into "families" of solutions, each family being characterized by the fact that for certain fixed indices $i_1,\ldots,i_p$ the corresponding variables have fixed values $x_{i_1} = \alpha_{i_1},\ldots,x_{i_p} = \alpha_{i_p}$, while the other variables: $x_{i_{p+1}},\ldots,x_{i_n}$, remain arbitrary. This way of expressing the solutions is very advantageous for our method of minimizing pseudo-Boolean functions (because that method uses certain pseudo-Boolean equations and inequalities; see part III).

In the present part we study only (systems of) linear equations and/or inequalities. The nonlinear case will be dealt with in part II.

The treatment of the linear case is, in the essence, a systematic accelerated search of the solutions within an associated tree. This tree-like construction was used by S.RUDEANU /33/ for solving Boolean equations; in the same paper, he conjuctured that the method can be adapted to the case of pseudo-

Boolean equations and inequalities*). This job will be done in the present part.

§ 2. Linear Pseudo-Boolean Equations

Let

(16) $$a_1z_1 + b_1\bar{z}_1 + a_2z_2 + b_2\bar{z}_2 + \ldots + a_nz_n + b_n\bar{z}_n = k,$$

where a_i, b_i $(i=1,\ldots,n)$ and k are given reals, be the general form of a linear pseudo-Boolean equation with the unknowns $z_1,\ldots,z_n$. Of course, we may assume, without loss of generality, that $a_i \neq b_i$ for all i (if not, the term $a_iz_i + b_i\bar{z}_i$ in the lefthand side of (16) is simply the constant a_i).

For each i, let us set

(17) $$x_i = \begin{cases} z_i, & \text{if } a_i > b_i, \\ \bar{z}_i, & \text{if } a_i < b_i. \end{cases}$$

Then the terms $a_iz_i + b_i\bar{z}_i$ may be transformed as follows:

(18) $$a_iz_i + b_i\bar{z}_i = \begin{cases} (a_i-b_i)\, x_i + b_i, & \text{if } a_i > b_i, \\ (b_i-a_i)\, x_i + a_i, & \text{if } a_i < b_i. \end{cases}$$

Thus, equation (16) becomes

(19) $$c_1x_1 + c_2x_2 + \ldots + c_nx_n = d\ ,$$

where $c_1,\ldots,c_n,d$ are reals, $c_i > 0$ $(i = 1,\ldots,n)$, and where

---

*) The SEP procedure for solving discrete extremum problems (B.ROY and B.SUSSMANN /32/, P.BERTIER and B.ROY /5/, P. BERTIER and Ph.T.NGHIEM /3/, P.BERTIER, Ph.T.NGHIEM and B.ROY /4/, B.ROY, Ph.T.NGHIEM and P.BERTIER /31/)is based on a similar idea. See also E.BALAŞ /1/.

(after re-indexing the unknowns), we can suppose that

(20) $$c_1 \geqslant c_2 \geqslant \ldots \geqslant c_n > 0.$$

Now, we are concentrating our attention on a procedure for solving the "canonical"form (19) under the assumption (20).

We shall track down the solutions of equation (19) along the branches of the tree in Fig. 1. This tree has n + 1 levels 0,1,...,n

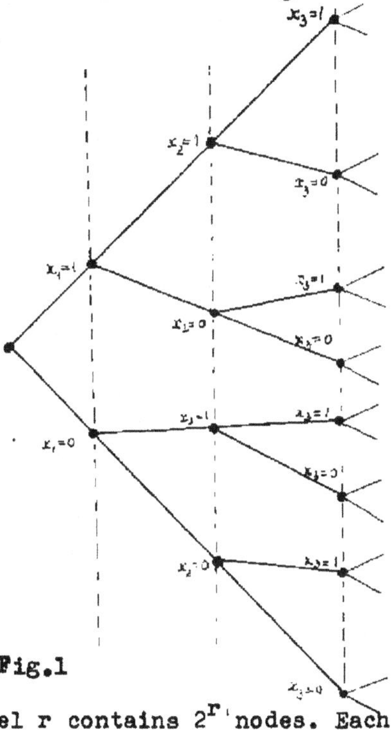

Fig.1

Each level r contains 2^r nodes. Each node of the r-th level is characterized by the fact that the values of the variables $x_1,\ldots,x_r$ are fixed ($x_1 = \xi_1,\ldots,x_r = \xi_r$), while the variables $x_{r+1},\ldots,x_n$ are subject to the equation

(19.r) $$\sum_{j=r+1}^{n} c_j x_j = d'$$

(where $d' = d - \sum_{k=1}^{r} c_k \xi_k$), which is of the type (19)).

Of course, it would be unreasonable to follow all the possible paths. Fortunately, most of the blind alleys can be avoided by a systematic use of the following:

- Table 2 -

No.	Case	Conclusions
1°.	$d<0$	No solutions
2°.	$d=0$	The unique solution is $x_1 = x_2 = \ldots = x_n = 0$
3°.	$d>0$ and $c_1 \geq \ldots \geq c_p > d \geq c_{p+1} \geq \ldots \geq c_n$	The solutions (if any) satisfy $x_1 = \ldots = x_p = 0$ and $\sum_{j=p+1}^{n} c_j x_j = d$
4°.	$d>0$ and $c_1 = \ldots = c_p = d > c_{p+1} \geq \ldots \geq c_n$	α) For every $k=1,2,\ldots,p$: $x_k=1$, $x_1=\ldots=x_{k-1}=x_{k+1}=\ldots=x_n=0$ is a solution. β) The other solutions (if any) satisfy $x_1=\ldots=x_p=0$ and $\sum_{j=p+1}^{n} c_j x_j = d$
5°.	$d>0$, $c_i<d (i=1,2,\ldots,n)$ and $\sum_{i=1}^{n} c_i < d$	No solutions
6°.	$d>0$, $c_i<d (i=1,2,\ldots,n)$ and $\sum_{i=1}^{n} c_i = d$	The unique solution is $x_1 = x_2 = \ldots = x_n = 1$
7°.	$d>0$, $c_i<d (i=1,2,\ldots,n)$ $\sum_{i=1}^{n} c_i > d$ and $\sum_{j=2}^{n} c_j < d$	The solutions (if any) satisfy $x_1 = 1$ and $\sum_{j=2}^{n} c_j x_j = d - c_1$
8°.	$d>0$, $c_i<d (i=1,2,\ldots,n)$ $\sum_{i=1}^{n} c_i > d$ and $\sum_{j=2}^{n} c_j \geq d$	The solutions (if any) satisfy either $x_1 = 1$, and $\sum_{j=2}^{n} c_j x_j = d - c_1$, or $x_1 = 0$ and $\sum_{j=2}^{n} c_j x_j = d$.

Table 2 discusses 8 mutually exclusive cases concerning equation (19) and covering all the situations; for each case, an obvious conclusion is drawn. We see that the following circumstances may occur:

- equation (19) is inconsistent (cases 1° and 5°);
- equation (19) has a unique solution (cases 2° and 6°);
- equation (19) is replaced by an equation of the same type, but with less variables (cases 3°, 4° and 7°);
- equation (19) is replaced by two equations of the same type, but with less variables; each of these equations is to be discussed separately (case 8°).

Therefore, unless equation (19) is inconsistent or it has a unique solution, we have to continue the investigation by applying the conclusions in Table 2 to the new equation(s) resulted at the first step. This process is continued until we have exhausted all the possibilities.

We have thus proved

THEOREM 3 (i) _The above described procedure leads to all the solutions of the canonical equation (19).(ii).If T is the transformation from (16) to (19), then the solutions of (16) are obtained by applying T^{-1} to the solutions of (19)._

Example 2. Let us solve the linear pseudo-Boolean equation

(21) $$4z_1 + \bar{z}_1 - 3z_2 + \bar{z}_2 + 5z_3 - 2z_4 + 5z_5 + 2z_6 - z_7 = 7.$$

Applying the transformation (17), we set

(22) $$y_1 = z_1,\ y_2 = \bar{z}_2,\ y_3 = z_3,\ y_4 = \bar{z}_4,\ y_5 = z_5,\ y_6 = z_6,\ y_7 = \bar{z}_7,$$

hence equation (26) becomes

$$(3y_1+1) + (4y_2-3) + 5y_3 + (2y_4-2) + 5y_5 + 2y_6 + (y_7-1) = 7,$$

or

$$3y_1 + 4y_2 + 5y_3 + 2y_4 + 5y_5 + 2y_6 + y_7 = 12,$$

or else, ordering the unknowns so as condition (20) be fulfilled, we get

(23) $$5x_1 + 5x_2 + 4x_3 + 3x_4 + 2x_5 + 2x_6 + x_7 = 12,$$

where

(24) $$\begin{cases} x_1 = y_3 = z_3 , \\ x_2 = y_5 = z_5 , \\ x_3 = y_2 = \bar{z}_2 , \\ x_4 = y_1 = z_1 , \\ x_5 = y_4 = \bar{z}_4 , \\ x_6 = y_6 = z_6 , \\ x_7 = y_7 = \bar{z}_7 . \end{cases}$$

We begin now the tree-like construction of the solutions.

The reader is adviced to follow simultaneously the computations and Fig.2.

Since we are in Case 8^0, we have simply to consider the two equations obtained from (23) by making $x_1=1$ and $x_1=0$. Equation (23) becomes

(23.1) $$5x_2 + 4x_3 + 3x_4 + 2x_5 + 2x_6 + x_7 = 7$$

and

(23.0) $$5x_2 + 4x_3 + 3x_4 + 2x_5 + 2x_6 + x_7 = 12$$

respectively. The number (23.1) indicates that the corresponding equation was obtained from (23) by making $x_1=1$; similarly for (23.0) etc. So, for instance, the label (23.1100) below indicates the equation obtained from (23) by making $x_1=1, x_2=1, x_3=0, x_4=0$.

We begin by following the branch $x_1 = 1$ corresponding to (23.1). We are again in Case 8^0, so that we continue the splitting with respect to x_2 :

(23.11) $$4x_3 + 3x_4 + 2x_5 + 2x_6 + x_7 = 2,$$

(23.10) $$4x_3 + 3x_4 + 2x_5 + 2x_6 + x_7 = 7.$$

Applying the conclusion of the case 3^0 to (23.11), we obtain $x_3 = x_4 = 0$ and the equation

(23.1100) $$2x_5 + 2x_6 + x_7 = 2;$$

by 4^0, we get the solutions

(23.1100100) $x_1=1,\ x_2=1,\ x_3=0,\ x_4=0,\ x_5=1,\ x_6=0,\ x_7=0,$

(23.1100010) $x_1=1,\ x_2=1,\ x_3=0,\ x_4=0,\ x_5=0,\ x_6=1,\ x_7=0,$

and the equation

(23.110000) $$x_7 = 2,$$

which has no solutions (by 5^0).

Now we come back to equation (23.10), which is in Case 8^0, so that we have to consider separately the cases $x_3 = 1$ and $x_3 = 0$:

(23.101) $$3x_4 + 2x_5 + 2x_6 + x_7 = 3,$$

(23.100) $$3x_4 + 2x_5 + 2x_6 + x_7 = 7.$$

Using the conclusion of the case 4^0, we see that equation (23.101) leads to the solution

(23.1011000) $x_1=1,\ x_2=0,\ x_3=1,\ x_4=1,\ x_5=0,\ x_6=0,\ x_7=0$

and to the equation

(23.1010) $$2x_5 + 2x_6 + x_7 = 3,$$

which, in its turn, leads to the equations

(23.10101) $$2x_6 + x_7 = 1,$$

(23.10100) $$2x_6 + x_7 = 3.$$

The conclusion 3°, applied to (23.10101), shows that $x_6 = 0$, hence $x_7 = 1$, therefore

(23.1010101) $x_1=1,\ x_2=0,\ x_3=1,\ x_4=0,\ x_5=1,\ x_6=0,\ x_7=1,$

while equation (23.10100) can be solved with the aid of 6° :

(23.1010011) $x_1=1,\ x_2=0,\ x_3=1,\ x_4=0,\ x_5=0,\ x_6=1,\ x_7=1.$

Now we come back to equation (23.100), which by a repeated application of the conclusion of the case 7°, implies first that $x_4 = 1$ and

(23.1001) $$2x_5 + 2x_6 + x_7 = 4,$$

then $x_5 = 1$ and

(23.10011) $$2x_6 + x_7 = 2,$$

hence, either by 4° and 5°, or by 7°, we deduce the solution

(23.1001110) $x_1=1,\ x_2=0,\ x_3=0,\ x_4=1,\ x_5=1,\ x_6=1,\ x_7=0.$

We have thus found all the solutions of equation(23.1), so that it remains to determine the solutions of (23.0).

We must split again, obtaining thus the equations

(23.01) $$4x_3 + 3x_4 + 2x_5 + 2x_6 + x_7 = 7,$$

(23.00) $$4x_3 + 3x_4 + 2x_5 + 2x_6 + x_7 = 12.$$

Equation (23.01) coincides with (23.10), whose solutions were determined before. Therefore, the solutions of (23.01) can be simply obtained from those of (23.10) by taking $x_1 = 0$ and $x_2 = 1$, instead of $x_1 = 1$ and $x_2 = 0$:

(23.0111000) $x_1=0,\ x_2=1,\ x_3=1,\ x_4=1,\ x_5=0,\ x_6=0,\ x_7=0,$

(23.0110101) $x_1=0,\ x_2=1,\ x_3=1,\ x_4=0,\ x_5=1,\ x_6=0,\ x_7=1,$

(23.0110011) $x_1=0,\ x_2=1,\ x_3=1,\ x_4=0,\ x_5=0,\ x_6=1,\ x_7=1,$

(23.0101110) $x_1=0,\ x_2=1,\ x_3=0,\ x_4=1,\ x_5=1,\ x_6=1,\ x_7=0.$

As to equation (23.00), 6^0 shows that it has the unique solution $x_3 = \ldots = x_7 = 1$, so that we obtain the following solution of initial equation (23):

(23.0011111) $x_1=0,\ x_2=0,\ x_3=1,\ x_4=1,\ x_5=1,\ x_6=1,\ x_7=1.$

We have thus found all the solution of (23); taking into account the transformation formulas (24), we obtain the solution of the given equation (21):

- Table 3 -

z_1	z_2	z_3	z_4	z_5	z_6	z_7
0	1	1	0	1	0	1
0	1	1	1	1	1	1
1	0	1	1	0	0	1
0	0	1	0	0	0	0
0	0	1	1	0	1	0
1	1	1	0	0	1	1
1	0	0	1	1	0	1
0	0	0	0	1	0	0
0	0	0	1	1	1	0
1	1	0	0	1	1	1
1	0	0	0	0	1	0

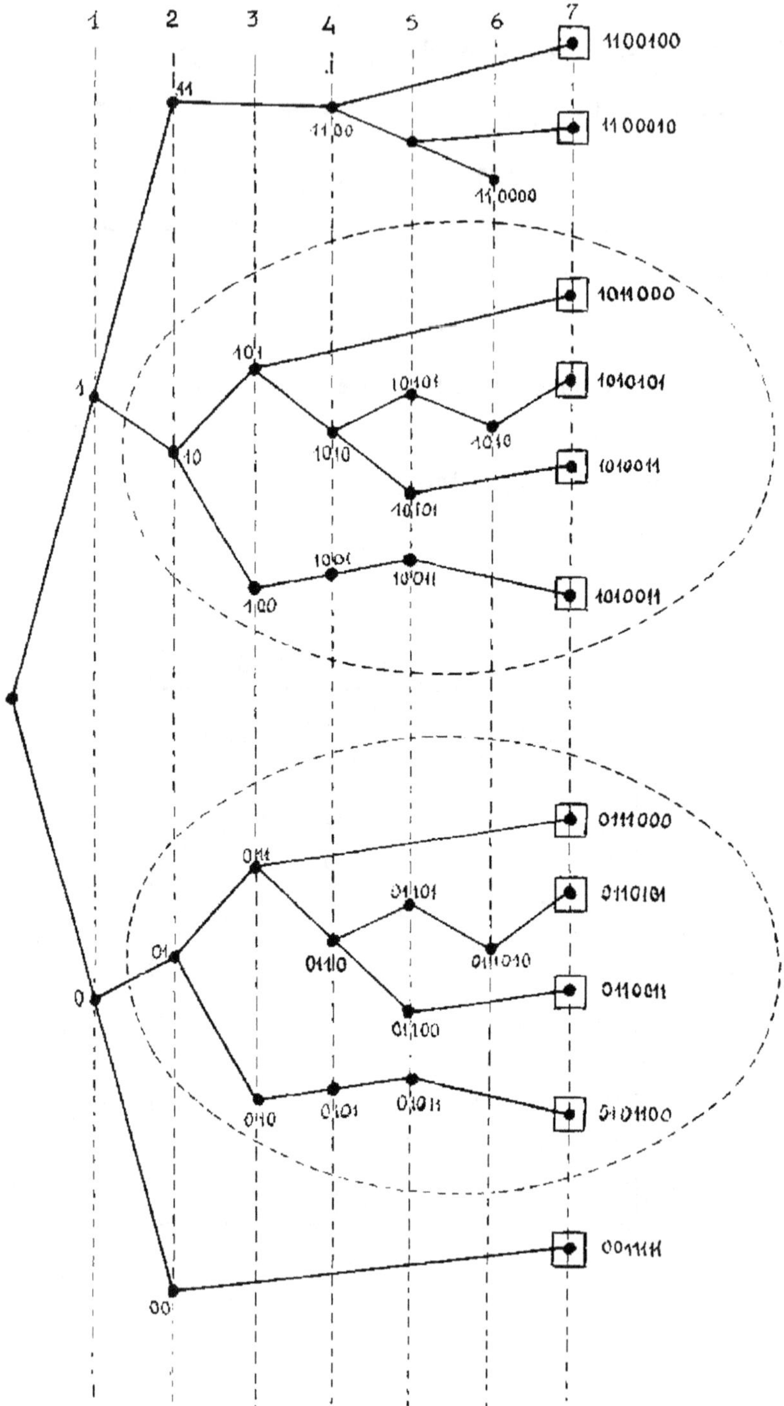
1
2
3
4
5
6
7
1100100
1100010
11,0000
1011000
1010101
1010011
1010011
0111000
0110101
0110011
0101100
001111

Fig. 2

Comments 1. The above described method determines all the solutions, and no solution is found twice.

2. In the above example, we have tested 12 paths, 11 of which have led to (all the) solutions. In other words, only one paths was unfruitful; the other 116 paths which correspond to non-solutions were avoided.

3. Moreover, the fact that we have obtained the same equation at two distinct stages of the proces (corresponding to the points 10 and 01) contribuited also to the reduction of the amount of còmputations.

§ 3. Linear Pseudo-Boolean Inequalities

The most general form of a linear pseudo-Boolean inequality is either

(25) $$a_1 z_1 + b_1 \bar{z}_1 + a_2 z_2 + b_2 \bar{z}_2 + \ldots + a_n z_n + b_n \bar{z}_n > h,$$

or

(26) $$a_1 z_1 + b_1 \bar{z}_1 + a_2 z_2 + b_2 \bar{z}_2 + \ldots + a_n z_n + b_n \bar{z}_n \geqslant k,$$

where a_i, b_i, h and k are reals and we may assume that $a_i \neq b_i$ for all i (if we have the sign $<$ or $\leqslant$ instead of $>$ or $\geqslant$, respectively, we multiply the whole inequality by - 1). In the common case when the coefficients a_i, b_i and h are integers, the strict inequality (25) may be also written in the form (26), if we take k = h + 1.

We shall confine our attention to inequalities of the form (26). As a matter of fact, the method developed in this section for solving the inequality (26), will directly offer the solutions of the equation

(16) $$a_1 z_1 + b_1 \bar{z}_1 + a_2 z_2 + b_2 \bar{z}_2 + \ldots + a_n z_n + b_n \bar{z}_n = k,$$

as well as those of the strict inequality

(27) $a_1z_1 + b_1\bar{z}_1 + a_2z_2 + b_2\bar{z}_2 + \ldots + a_nz_n + b_n\bar{z}_n > k,$

in case that a_i, b_i and k are integers.

We shall prove that the solutions of the inequality (26) (if any) can be grouped into "families of solutions", in the sense of the following

Definition 2. Let $S = (z_1^*, \ldots, z_n^*)$ be a solution of (26) and let I be a set of indices: $I \subseteq \{1,2,\ldots,n\}$. Let $\sum(S,I)$ be the set of all vectors $(z_1,\ldots,z_n) \in B_2^n$ satisfying

$$z_i = z_i^* \text{ for all } i \in I,$$

the other variables z_j $(j \notin I)$ being arbitrary. If all the vectors $(z_1,\ldots,z_n) \in \sum(S,I)$ satisfy the inequality (26), then $\sum(S,I)$ is said to be a family of solutions of (26). We say also that this family is generated by the pair (S,I); the variables z_i for which $i \in I$ are called the fixed variables of the family.

Notice that relation $S \in \sum(S,I)$ holds for every pair (S,I). If $I = \{1,2,..,n\}$, then $\sum(S,I)$ is a degenerate family containing a single solution, namely S. More generally, if the set I consists of r indices, then $\sum(S,I)$ contains 2^{n-r} elements; if $r < n$, the family may be called "non-degenerate".

We want to obtain the solutions grouped into families of solutions, so that the number of these families should be as small as possible. Therefore we are interested in obtaining, whenever possible, non-degenerate families of solutions. Moreover, it will be shown that it is possible to obtain the solutions grouped into set-theoretically disjoint families.

Since it is not easy to detect non-degenerate families

of solutions directly on the inequality (26), we shall first reduce it to a standard form. Namely, applying to each z_i the transformation (17) and re-ordering the unknowns, as in § 1, we see that the inequality (26) may be brought to the canonical form

(28) $$c_1x_1 + c_2x_2 + \ldots + c_nx_n \geqslant d,$$

where $c_1,\ldots,c_n$, d are reals and

(29) $$c_1 \geqslant c_2 \geqslant \ldots \geqslant c_n > 0.$$

We give below a procedure which enables us to obtain the solutions of (28) grouped into several non-degenerate and pairwise disjoint families of solutions; after this has been done, we apply the inverse transformation (from (28)to (26)) and obtain immediately the families of solutions of (26).

To this end, let us introduce the following

Definition 3. A vector $(x_1^*,\ldots,x_n^*)$ satisfying the inequality (28) is called a basic solution of (28), if for each index i such that $x_i^* = 1$, the vector $(x_1^*,\ldots,x_{i-1}^*, 0, x_{i+1}^*,\ldots,x_n^*)$ is not a solution of (28).

Remark 1. The solutions of the equation

(19) $$c_1x_1 + c_2x_2 + \ldots + c_nx_n = d$$

(if any) are basic solution of the inequality (28).

We shall prove that the solutions of (28) may be found by a process involving two steps:

a) Determine all the basic solutions of (28).

b) To each basic solution S_k associate a certain set of indices I_k in such a way that $\sum(S_k, I_k)$ should be a family of solutions and that the system $\left\{\sum (S_k, I_k)\right\}_{k=1,\ldots,m}$ should be "complete" (i.e.,it should include all the solutions of (28)).

We proced now to the first step:

a) Determination of the basic solutions

The basic solutions of (28) will be determined by a tree-like construction similar to that used for solving linear equations.

The following three lemmas are easy to prove:

LEMMA 1. Let $(x_1^*,\ldots,x_p^*, x_{p+1}^*,\ldots,x_n^*)$ be a basic solution of (28); then $(x_{p+1}^*,\ldots,x_n^*)$ is a basic solution of the inequality

$$\sum_{j=p+1}^{n} c_j x_j \geq d - \sum_{k=1}^{p} c_k x_k^* \,. \tag{30}$$

LEMMA 2. If $(x_{p+1}^*,\ldots,x_n^*)$ is a basic solution of the inequality

$$\sum_{j=p+1}^{n} c_j x_j \geq d, \tag{31}$$

then $(\underbrace{0,\ldots,0}_{p \text{ times}}, x_{p+1}^*,\ldots,x_n^*)$ is a basic solution of (28).

LEMMA 3. If $(x_2^*,\ldots,x_n^*)$ is a basic solution of

$$\sum_{j=2}^{n} c_j x_j \geq d - c_1, \tag{32}$$

then $(1,x_2^*,\ldots,x_n^*)$ is a basic solution of (28).

Lemmas 1,2 and 3 enable us be build up the following Table 4, which is the analogue of Table 2 in § 1.

- Table 4 -

No	Case	Conclusions	Valid
1°	$d \leq 0$	The unique basic solution is $x_1 = x_2 = \ldots = x_n = 0$	Obviously
2°	$d > 0$ and $c_1 \geq \ldots \geq c_p \geq d > c_{p+1} \geq \ldots \geq c_n$	α) For every $k = 1,2,\ldots,p$: $x_k=1, x_1=\ldots=x_{k-1}=x_{k+1}=\ldots=x_n=0$ is a basic solution .	Obviously
		β) The other basic solutions (if any) are characterized by the property: $x_1=\ldots=x_p = 0$, and $(x_{p+1},\ldots,x_n)$ is a basic solution of $$\sum_{j=p+1}^{n} c_j x_j \geq d$$	by lemmas 3,4
3°	$d>0, c_i<d (i=1,2,\ldots,n)$ and $\sum_{i=1}^{n} c_i < d$	No solutions	Obviously
4°	$d>0, c_i<d (i=1,2,\ldots,n)$ and $\sum_{i=1}^{n} c_i = d$	The unique (basic) solution is $x_1 = x_2 = \ldots = x_n = 1$	Obviously
5°	$d>0, c_i<d (i=1,2,\ldots,n)$ $\sum_{i=1}^{n} c_i > d$ and $\sum_{j=2}^{n} c_j < d$	The basic solutions (if any) are characterized by the property: $x_1 = 1$, and $(x_2,\ldots,x_n)$ is a basic solution of $\sum_{j=2}^{n} c_j x_j \geq d - c_1$	by lemmas 3,5
6°	$d>0, c_i<d (i=1,2,\ldots,n)$ $\sum_{i=1}^{n} c_i > d$ and $\sum_{j=2}^{n} c_j \geq d$	The basic solution (if any) are characterized by the property : either $x_1 = 1$ and $(x_2,\ldots,x_n)$ is a basic solution of $$\sum_{j=2}^{n} c_j x_j \geq d - c_1$$ or : $x_1=0$ and $(x_2,\ldots,x_n)$ is a basic solution of $$\sum_{j=2}^{n} c_j x_j \geq d$$	by lemmas 3,5 and 4

We conclude:

THEOREM 4. <u>The above described method determines all the basic solutions of (28)</u>.

We proceed now to the step

b) <u>Determination of a complete system of families of solutions of (28)</u>

To each basic solution $S = (x_1^*, \ldots, x_n^*)$ we associate a family of solutions $\sum(S, J_S)$ defined as follows. Let i_0 be the last index for which $x_i^* = 1$, (i.e. $x_{i_0}^* = 1$ and $x_i^* = 0$ for all $i > i_0$), and let J_S be the set of all indices $i \leq i_0$. Then $\sum(S, J_S)$ (see Definition 2) is the set of all vectors $(x_1, \ldots \ldots x_n)$ satisfying

(33) $$x_i = \begin{cases} x_i^*, & \text{for } i \leq i_0 \\ \text{arbitrary}, & \text{for } i > i_0 . \end{cases}$$

Using the above results, we can prove the following :

THEOREM 5. <u>Let</u> $S_1, \ldots, S_m$ <u>be all the basic solutions of (28) and let</u> $\sum_k = \sum(S_k, J_{S_k})$ $(k=1,\ldots,m)$ <u>be constructed as above. Then every solution</u> $(x_1, \ldots, x_n)$ <u>of (28) belongs to exactly one of these families of solutions</u>.

Concluding this discussion, we come to

THEOREM 6. <u>The procedure summarized in Theorems 4 and 5 gives all the solutions of the canonical inequality (28)</u>.

<u>Corollary 1</u>. Let T be the transformation leading from (26) to (28); we apply the inverse transformation T^{-1} to the solutions of (28) and obtain the solutions of the original inequality (28).

<u>Example 3</u>. Let us solve the linear pseudo-Boolean inequality

(34) $$2\bar{z}_1 - 5z_2 + 3z_3 + 4\bar{z}_4 - 7z_5 + 16z_6 - z_7 \geqslant -4.$$

We set

(35) $$z_1=\bar{y}_1,\ z_2=\bar{y}_2,\ z_3=y_3,\ z_4=\bar{y}_4,\ z_5=\bar{y}_5,\ z_6=y_6,\ z_7=\bar{y}_7\ ,$$

hence

$$2y_1 + 5y_2 + 3y_3 + 4y_4 + 7y_5 + 16y_6 + y_7 \geqslant 9$$

or else

(36) $$16x_1 + 7x_2 + 5x_3 + 4x_4 + 3x_5 + 2x_6 + x_7 \geqslant 9,$$

where

(37) $$\begin{cases} x_1 = y_6 = z_6 \\ x_2 = y_5 = \bar{z}_5 \\ x_3 = y_2 = \bar{z}_2 \\ x_4 = y_4 = \bar{z}_4 \\ x_5 = y_3 = z_3 \\ x_6 = y_1 = \bar{z}_1 \\ x_7 = y_7 = \bar{z}_7 \end{cases}$$

The first coefficient being >9, we apply 2° and obtain the basic solution

(36.1000000) $x_1=1,\ x_2=0,\ x_3=0,\ x_4=0,\ x_5=0,\ x_6=0,\ x_7=0;$

the other basic solutions satisfy $x_1 = 0$ and

(36.0) $$7x_2 + 5x_3 + 4x_4 + 3x_5 + 2x_6 + x_7 \geqslant 9$$

(here again, as in the previous section, the label (36.0) indicates the inequality obtained from (36) by making $x_1 = 0$).

As we are now in case 6°, we shall examine distincly the inequalities

(36.01) $$5x_3 + 4x_4 + 3x_5 + 2x_6 + x_7 \geqslant 2,$$

(36.00) $$5x_3 + 4x_4 + 3x_5 + 2x_6 + x_7 \geqslant 9,$$

corresponding to $x_2 = 1$ and $x_2 = 0$, respectively.

Applying now 2° to (36.01), we obtain the following basic solutions:

(36.0110000) $x_1=0,\ x_2=1,\ x_3=1,\ x_4=0,\ x_5=0,\ x_6=0,\ x_7=0,$

(36.0101000) $x_1=0,\ x_2=1,\ x_3=0,\ x_4=1,\ x_5=0,\ x_6=0,\ x_7=0,$

(36.0100100) $x_1=0,\ x_2=1,\ x_3=0,\ x_4=0,\ x_5=1,\ x_6=0,\ x_7=0,$

(36.0100010) $x_1=0,\ x_2=1,\ x_3=0,\ x_4=0,\ x_5=0,\ x_6=1,\ x_7=0,$

and the inequality

(36.010000) $$x_7 \geqslant 2,$$

which has no solutions (see 3°).

We come back now to the inequality (36.00), which satisfies 6°. We consider the two subcases $x_3 = 1$ and $x_3 = 0$:

(36.001) $$4x_4 + 3x_5 + 2x_6 + x_7 \geqslant 4,$$

(36.000) $$4x_4 + 3x_5 + 2x_6 + x_7 \geqslant 9.$$

The inequality (36.001) has the basic solution (see 2°) $x_4=1,\ x_5 = x_6 = x_7 = 0$, leading to

(36.0011000) $x_1=0,\ x_2=0,\ x_3=1,\ x_4=1,\ x_5=0,\ x_6=0,\ x_7=0,$

while the other basic solutions satisfy $x_4 = 0$ and

(36.0010) $$3x_5 + 2x_6 + x_7 \geqslant 4;$$

the conclusion 5° shows that $x_5 = 1$ and

(36.00101) $$2x_6 + x_7 \geqslant 1,$$

which, in view of 2°, admits the basic solutions $x_6=1,\ x_7=0$ and $x_6=0,\ x_7=1$, leading to:

(36.0010110) $x_1=0,\ x_2=0,\ x_3=1,\ x_4=0,\ x_5=1,\ x_6=1,\ x_7=0,$

and to

(36.0010101) $x_1=0,\ x_2=0,\ x_3=1,\ x_4=0,\ x_5=1,\ x_6=0,\ x_7=1,$

respectively.

We have to consider the inequality (36.000) which falling into the case 5°, implies $x_4 = 1$ and

(36.0001) $$3x_5 + 2x_6 + x_7 \geq 5.$$

This inequality, by the same argument, gives $x_5 = 1$ and

(36.00011) $$2x_6 + x_7 \geq 2;$$

then, case 2° shows that we have the basic solutions

(36.0001110) $x_1=0,\ x_2=0,\ x_3=0,\ x_4=1,\ x_5=1,\ x_6=1,\ x_7=0$

and the inequality

(36.000110) $$x_7 \geq 2,$$

which has no solutions (by 3°).

The tree-like construction having come to an end, we have obtained all the basic solutions of (36), which we group together in Table 5 below, where we indicate by a label the solutions of the equation

(38) $$16x_1 + 7x_2 + 5x_3 + 4x_4 + 3x_5 + 2x_6 + x_7 = 9$$

(see Remark 1).

- Table 5 -

No.	x_1^*	x_2^*	x_3^*	x_4^*	x_5^*	x_6^*	x_7^*	(38)?
1	1	0	0	0	0	0	0	
2	0	1	1	0	0	0	0	
3	0	1	0	1	0	0	0	
4	0	1	0	0	1	0	0	
5	0	1	0	0	0	1	0	✓
6	0	0	1	1	0	0	0	✓
7	0	0	1	0	1	1	0	
8	0	0	1	0	1	0	1	✓
9	0	0	0	1	1	1	0	✓

The corresponding families of solutions $\sum(S,J_S)$ of (36) are given in Table 6 below, where the dashes indicate the arbitrary variables.

- Table 6 -

No.	x_1	x_2	x_3	x_4	x_5	x_6	x_7
1	1	-	-	-	-	-	-
2	0	1	1	-	-	-	-
3	0	1	0	1	-	-	-
4	0	1	0	0	1	-	-
5	0	1	0	0	0	1	-
6	0	0	1	1	-	-	-
7	0	0	1	0	1	1	-
8	0	0	1	0	1	0	1
9	0	0	0	1	1	1	-

Taking into account the transformation formulas (37), we obtain the families of solutions of the original inequality (34):

- Table 7 -

No.	z_1	z_2	z_3	z_4	z_5	z_6	z_7
1	-	-	-	-	-	1	-
2	-	0	-	-	0	0	-
3	-	1	-	0	0	0	-
4	-	1	1	1	0	0	-
5	0	1	0	1	0	0	-
6	-	0	-	0	1	0	-
7	0	0	1	1	1	0	-
8	1	0	1	1	1	0	0
9	0	1	1	0	1	0	-

As we have announced at the beginning of this section, in case of inequalities with integer coefficients we can solve simultaneously the inequality (26), the equation (16) and the strict inequality (27). It suffices to know how to solve simultaneously the inequality (28), the strict inequality

(39) $$c_1x_1 + c_2x_2 + \dots + c_nx_n > d,$$

and the equation

(40) $$c_1x_1 + c_2x_2 + \dots + c_nx_n = d.$$

We have already noticed (Remark 1) that the solutions of (40) are to be sought among the basic solutions of (28).

The knowledge of the families of solutions of the inequality (28) enables us to determine the families of solutions of the strict inequality (39).

For inequalities of the type (28), the determination of the families of solutions reduces to the finding of the basic solutions. It turns out that the inequality (39) is actually of the type (28), since it may be written in the form

(39') $$c_1x_1 + c_2x_2 + \dots + c_nx_n \geqslant d + 1,$$

so that we have to find its basic solutions.

To do this, we examine the basic solutions of (28), which satisfy either (i) the strict inequality (39), or (ii) the equation (40). The solutions (i) are obviously basic solutions of the strict inequality (39). As to the case (ii), consider a solution $(x_1^*,\dots,x_n^*)$ of the equation (28), and let p be the place of the last 1 in this solution i.e. : $x_p^* = 1$, $x_{p+1}^* = \dots = x_n^* = 0$. We change, in turn, each of the last n-p zeros into 1, obtaining thus n-p vectors which will prove to be basic solutions of the strict inequality (39).

The above described procedure provides us with all the basic solutions of (39). More exactly, the procedure runs as follows.

Let B be the set of all the basic solutions of (28). Let M' be the set of those basic solutions of (28) which are not solutions of the equation (40). Let now $S^* = (x_1^*,\dots,x_n^*)$ be an element of B-M' (i.e., a solution of (40)) and let p be the greatest index for which $x_p^* = 1$. We associate to S^* the vectors $R_j^* = (y_{j1}^*,\dots,y_{jn}^*)$ $(j = p+1,\dots,n)$ defined as follows

(41) $$y_{ji}^* = \begin{cases} x_i^* & \text{if } i \neq j, \\ 1 = \overline{x_j^*}, & \text{if } i = j. \end{cases}$$

The set of all the vectors R_j^* $(j = p+1,\dots,n)$ associated to the different elements of B-M' will be denoted by M".

Let us denote by M the set of all basic solutions of (39). Then we have:

THEOREM 7. Assume that $c_1,\dots,c_n$, d are integers. Then $M = M' \cup M''$.

Corollary 2. The solutions of the strict inequality (39) may be determined as follows: a) Find the basic solutions of (39) as indicated by Theorem 7. b) Find the families of solutions as indicated by Theorem 5.

Example 4. In Example 3 we have solved the inequality

(34) $$2\bar{z}_1 - 5z_2 + 3z_3 + 4\bar{z}_4 - 7z_5 + 16z_6 - z_7 \geqslant -4,$$

which has the canonical form

(36) $$16x_1 + 7x_2 + 5x_3 + 4x_4 + 3x_5 + 2x_6 + x_7 \geqslant 9,$$

the basic solutions of which were given in Table 5.

Table 8 below gives the solutions of the equation

(42) $$2\bar{z}_1 - 5z_2 + 3z_3 + 4\bar{z}_4 - 7z_5 + 16z_6 - z_7 = -4$$

associated to the inequality (34); they are simply the transforms of the solutions of (38), which were labelled in Table 5.

- Table 8 -

No.	z_1	z_2	z_3	z_4	z_5	z_6	z_7
5	0	1	0	1	0	0	1
6	1	0	0	0	1	0	1
8	1	0	1	1	1	0	0
9	0	1	1	0	1	0	1

Further, in order to solve the strict inequality

(43) $$2\bar{z}_1 - 5z_2 + 3z_3 + 4\bar{z}_4 - 7z_5 + 16z_6 - z_7 > -4 ,$$

we have to find the basic solutions of the canonical strict inequality

(44) $$16x_1 + 7x_2 + 5x_3 + 4x_4 + 3x_5 + 2x_6 + x_7 > 9.$$

According to Theorem 7, these solutions are: (i) those basic solutions of (36) which do not satisfy (38) (these solutions are simply the non-labelles solutions in Table 5); (ii) the solutions of (44) associated to the solutions of (38) (the latter are the labelled solutions in Table 5). We obtain thus the table of all the basic solutions of the inequality (44) :

- Table 8 -

No.	x_1^*	x_2^*	x_3^*	x_4^*	x_5^*	x_6^*	x_7^*
1	1	0	0	0	0	0	0
2	0	1	1	0	0	0	0
3	0	1	0	1	0	0	0
4	0	1	0	0	1	0	0
5'	0	1	0	0	0	1	1
6'	0	0	1	1	1	0	0
6"	0	0	1	1	0	1	0
6"'	0	0	1	1	0	0	1
7	0	0	1	0	1	1	0
9'	0	0	0	1	1	1	1

In Tables 5 and 8, we have denoted by the same number a solution of (38) and the associated solution of (44). Notice that there is no solution of (44) associated to the solution No.8.

From the basic solutions in Table 8 we obtain the families of solutions of (44), and then, those of (43), by the same procedure as in Example 3. The result is given in Table 9,

- Table 9 -

No.	z_1	z_2	z_3	z_4	z_5	z_6	z_7
1	-	-	-	-	-	1	-
2	-	0	-	-	0	0	-
3	-	1	-	0	0	0	-
4	-	1	1	1	0	0	-
5'	0	1	0	1	0	0	0
6'	-	0	1	0	1	0	-
6"	0	0	0	0	1	0	-
6"'	1	0	0	0	1	0	0
7	0	0	1	1	1	0	-
9'	0	1	1	0	1	0	0

The families 1,2,3,4 and 7 are, of course, the same as in Table 7 while the families 5,6 and 9 in that table were replaced by the families 5',6',6",6"' and 9' generated by the associated solutions of (44). Although Table 9 contains no family corresponding to the family 8 of Table 7, Theorem 7 assures us that the solutions of (43) which belong to the family 8 in Table 7 are not lost: they are contained in various families of Table 9.

§ 4. Systems of Linear Pseudo-Boolean Equations and/or Inequalities

The method exposed in the preceding two sections for solving a linear pseudo-Boolean equation or inequality can by easily adapted to the more general case of a system of linear equations and/or inequalities.

The algorithm for solving linear systems will comprise three stages.

Stage 1. Replacing each inequality $h \leq 0$ by $-h \geq 0$, we obtain a system containing inequalities of the form $F \geq 0$, or equations $G = 0$, or both. In case of integer coefficients, inequalities of the form $f > 0$ (g 0) can also be dealt with, by replacing them by $f - 1 \geq 0$ ($-g - 1 \geq 0$).

Stage 2. Let $x_1, \ldots, x_n$ be the unknowns of the system. Using the relations $\bar{x}_i = 1 - x_i$ and/or $x_j = 1 - \bar{x}_j$, we can write, for each i, the i-th inequality, in the form

$$(45) \qquad c^i_{i_1} \tilde{x}_{i_1} + c^i_{i_2} \tilde{x}_{i_2} + \ldots + c^i_{i_m} \tilde{x}_{i_m} \geq d^i,$$

where: $x_{i_1}, \ldots, x_{i_m}$ are those variables the corresponding inequality depends effectively on, $\tilde{x}$ is either x or $\bar{x}$, so that $c^i_{i_1} \geq c^i_{i_2} \geq \ldots \geq c^i_{i_m} > 0$. The equations of the system are to be written in a similar way. In other words, we bring each equation and inequality to the canonical form with respect to the variables occuring effectively in it, but without changing the notation.

Stage 3. We apply now the following idea. Each equation (inequality), considered separately, is written in the canonical form with respect to the variables $\tilde{x}$ contained in it, therefore a certain conclusion can be drawn from Table 2 (respectively, from Table 4); this deduction leads to another conclusion referring to the whole system.

For instance, when a certain inequality or equation of the system has no solutions, then the whole system is inconsistent. In the same way, if the equation $f(x_{i_1},\dots,x_{i_m})=0$ has the unique solution $x_{i_1} = x^*_{i_1},\dots,x_{i_m} = x^*_{i_m}$, then each solution of the system (if any!) must satisfy the remaining relations, the variables $x_{i_1},\dots,x_{i_m}$ (which are not necessarily exhausting the set of all the variables of the system) having the above fixed values.

Further, we cannot transpose the notion of basic solution to the case of a system of linear inequalities; therefore, the conclusions in Table 4 are to be re-formulated so as to indicate the corresponding families of solutions. For instance, assume that the inequality (15) is in the case 2, that is $c^i_{i_1} \geqslant c^i_{i_2} \geqslant \dots \geqslant c^i_{i_p} > d^i \geqslant c^i_{i_{p+1}} \geqslant \dots \geqslant c_{i_m}$. Then, instead of the basic solutions

(46.k) $\widetilde{x}_{i_k} = 1,\ \widetilde{x}_{i_1} = \dots = \widetilde{x}_{i_{k-1}} = \widetilde{x}_{i_{k+1}} = \dots = \widetilde{x}_{i_m} = 0$ $(k=1,\dots,p)$

we have to consider simply the p branches

(47.k) $\widetilde{x}_{i_k} = 1,\ \widetilde{x}_{i_j}$ arbitrary for $j \neq k$ $(k=1,\dots,p)$.

Of course, it is convenient to consider set-theoretically disjoint families of solutions, so that we shall follow the branches

(48.k) $\widetilde{x}_{i_1} = \dots = \widetilde{x}_{i_{k-1}} = 0,\ \widetilde{x}_{i_k} = 1,\ \widetilde{x}_{i_j}$ arbitrary for $j > k$ $(k=1,\dots,p)$,

insted of (47.k).

We give below the complete list of these conclusions.

- Table 10 -

A. Equation

No.	C a s e	I n f o r m a t i o n s		
		Conclusions	Fixed variables	Remaining equation
1°	$d^i < 0$	No solutions	-	-
2°	$d^i = 0$	All of appearing variables fixed	$\tilde{x}_{i_1} = \dots = \tilde{x}_{i_m} = 0$	-
3°	$d^i > 0$ and $c^i_{i_1} \geqslant \dots c^i_{i_p} > d^i \geqslant$ $\geqslant c^i_{i_{p+1}} \geqslant \dots \geqslant c^i_{i_m}$	Part of appearing variables fixed	$\tilde{x}_{i_1} = \dots = \tilde{x}_{i_p} = 0$	$\sum_{j=p+1}^{m} c^i_{i_j} \tilde{x}_{i_j} = d^i$
4°	$d^i > 0$ and $c^i_{i_1} = \dots = c^i_{i_p} = d^i >$ $> c^i_{i_{p+1}} \geqslant \dots \geqslant c^i_{i_m}$	There are p+1 possibilities $\alpha_1, \dots, \alpha_p, \beta$	α_k: $\tilde{x}_{i_k} = 1$, $\tilde{x}_{i_1} = \dots = \tilde{x}_{i_{k-1}} = \tilde{x}_{i_{k+1}} = \dots = \tilde{x}_{i_m} = 0$ $(k=1, \dots, p)$	-
			β: $x_{i_1} = \dots = x_{i_p} = 0$	$\sum_{j=p+1}^{m} c^i_{i_j} \tilde{x}_{i_j} = d^i$
5°	$d^i > 0, c^i < d^i (j=1,2,\dots,m)$ and $\sum_{j=1}^{m} c^i_{i_j} < d^i$	No solutions	-	-
6°	$d^i > 0, c^i_{i_j} < d^i (j=1,2,\dots,m)$ and $\sum_{j=1}^{m} c^i_{i_j} = d^i$	All of appearing variables fixed	$\tilde{x}_{i_1} = \dots = \tilde{x}_{i_m} = 1$	-
7°	$d^i > 0, c^i_{i_j} < d^i (j=1,2,\dots,m)$ $\sum_{j=1}^{m} c^i_{i_j} > d^i$ and $\sum_{j=2}^{m} c^i_{i_j} < d^i$	One variable fixed	$\tilde{x}_{i_1} = 1$	$\sum_{j=2}^{m} c^i_{i_j} \tilde{x}_{i_j} = d^i - c^i_{i_1}$
8°	$d^i > 0, c^i_{i_j} < d^i (j=1,2,\dots,m)$ $\sum_{j=1}^{m} c^i_{i_j} > d^i$ and $\sum_{j=2}^{m} c^i_{i_j} \geqslant d^i$	There are two possibilities γ_1, γ_2	γ_1 : $\tilde{x}_{i_1} = 1$	$\sum_{j=2}^{m} c^i_{i_j} \tilde{x}_{i_j} = d^i - c^i_{i_1}$
			γ_2 : $\tilde{x}_{i_1} = 0$	$\sum_{j=2}^{m} c^i_{i_j} \tilde{x}_{i_j} = d^i$

- Table 10 -

B. Inequality

No.	Case	Informations		
		Conclusions	Fixed variables	Remaining inequality
1°	$d^i \leq 0$	Redundant inequality	-	-
2°	$d^i > 0$ and $c^i_{i_1} \geq \dots c^i_{i_p} \geq d^i >$ $> c^i_{i_{p+1}} \geq \dots \geq c^i_{i_m}$	There are p+1 possibilities $\alpha_1, \dots, \alpha_p, \beta$	α_k: $\tilde{x}_{i_1} = \dots = \tilde{x}_{i_{k-1}} = 0$, $\tilde{x}_{i_k} = 1$ (k=1,...,p)	-
			β: $x_{i_1} = \dots = x_{i_p} = 0$	$\sum_{j=p+1}^{m} c^i_{i_j} \tilde{x}_{i_j} \geq d^i$
3°	$d^i > 0$, $c^i_{i_j} < d^i$ (j=1,...,m) and $\sum_{j=1}^{m} c^i_{i_j} < d^i$	No solutions	-	-
4°	$d^i > 0$, $c^i_{i_j} < d^i$ (j=1,...,m) $\sum_{j=1}^{m} c^i_{i_j} = d^i$	All of appearing variables fixed	$\tilde{x}_{i_1} = \dots = \tilde{x}_{i_m} = 1$	-
5°	$d^i > 0$, $c^i_{i_j} < d^i$ (j=1,...,m) $\sum_{j=1}^{m} c^i_{i_j} > d^i$ and $\sum_{j=2}^{m} c^i_{i_j} < d^i$	One variable fixed	$\tilde{x}_{i_1} = 1$	$\sum_{j=2}^{m} c^i_{i_j} \tilde{x}_{i_j} \geq d^i - c^i_{i_1}$
6°	$d^i > 0$, $c^i_{i_j} < d^i$ (j=1,2,...,m) $\sum_{j=1}^{m} c^i_{i_j} > d^i$ and $\sum_{j=2}^{m} c^i_{i_j} \geq d^i$	There are two possibilities γ_1, γ_2	γ_1 : $\tilde{x}_{i_1} = 1$	$\sum_{j=2}^{m} c^i_{i_j} \tilde{x}_{i_j} \geq d^i - c^i_{i_1}$
			γ_2 : $\tilde{x}_{i_1} = 0$	$\sum_{j=2}^{m} c^i_{i_j} \tilde{x}_{i_j} \geq d^i$

As we see, there are cases in which some variables are fixed, or in which there are no solutions,or in which the considered equation (inequality) is redundant;we call these cases "determinate". There are other cases when we have practically no informations and we are obliged to split the discussion into two cases; we call these cases "undeterminate". Finally, there are cases when the discussion is to be splitted into p+1 cases with increased informations; we call them "partially determinate". This classification is given in Table 11 below:

- Table 11 -

Preferential order	Equation (Table 10 A)	Inequality (Table 10 B)	Characterization
First	1°, 5° 2°, 6° 3°, 7°	3° 1°, 4° 5°	"Determinate"
Second	4°	2°	"Partially determinate"
Third	8°	6°	"Undeterminate"

Now the 3-th stage of the procedure for solving a system of linear equations and/or inequalities may be continued as follows:

If some equations and inequalities belong to "determinate" cases we draw all the corresponding conclusions and collate them. Two situations may arrise. If at least one equation or inequality has no solutions, or if two distinct equations or inequalities lead to conclusions of the form $x_i=1$ and $x_i=0$ respectively, then the system has no solutions. In the other cases, the values of certain variables are determined and this leads to a smaller system which is to be examined.

If none of the equations and inequalities is in a determinate case, but there are equations or inequalities in partially determinate cases, then we follow the conclusion corresponding to one of these cases. It seems convenient to choose the equation or inequality corresponding to the greatest p (see Table 10, the cases 4^{0}A and 2^{0}B).

Finally, if all the equations and inequalities are in undeterminate cases, we split the discussion with respect to one of the variables; it seems convenient to choose (one of) the variable(s) appearing with the greatest coefficient in the system.

We have the following

THEOREM 8. <u>The above described procedure leads to all the solutions of the considered system of linear pseudo-Boolean equations and/or inequalities.</u>

Of course, the above method may be enriched by adding several supplementary rules for speeding up the computations. In Example 5 below, we have deliberately abstained from using such accelerating remarks, in order to illustrate only the essence of the procedure.

Remark 2. The families of solutions obtained by the method described in this section corresponding to different branches of the associated tree (Fig.1), so that these families are pairwise disjoint.

Example 5. Let us solve the system

(49.1) $2x_1 - 4x_2 + 8x_3 + 3x_4 - 6x_5 = -2,$

(49.2) $5x_1 - 4x_3 + 3x_5 + 2x_6 - x_7 + 9x_8 \leq 5,$

(49.3) $4x_1 + 6x_2 + 4x_4 - 5x_5 - 9x_6 + 8x_7 > -1,$

(49.4) $2x_2 - 4x_4 - x_6 + 3x_8 \geqslant 1.$

Performing the transformations indicated at the stages 1 and 2, we obtain the following equivalent system:

(50.1) $8x_3 + 6\bar{x}_5 + 4\bar{x}_2 + 3x_4 + 2x_1 = 8,$

(50.2) $9\bar{x}_8 + 5\bar{x}_1 + 4x_3 + 3\bar{x}_5 + 2\bar{x}_6 + x_7 \geqslant 14,$

(50.3) $9\bar{x}_6 + 8x_7 + 6x_2 + 5\bar{x}_5 + 4x_1 + 4x_4 \geqslant 14,$

(50.4) $4\bar{x}_4 + 3x_8 + 2x_2 + \bar{x}_6 \geqslant 6.$

Equation (50.1) is in the "partially determinate" case 4°A, while the other relations are in "undeterminate" cases. Here we have $p = 1$, therefore there are two alternatives:

α) $x_3=1,\ \bar{x}_5 = \bar{x}_2 = x_4 = x_1 = 0$

and

β) $x_3=0$ and $6\bar{x}_5 + 4\bar{x}_2 + 3x_4 + 2x_1 = 8.$

In the alternative α, equation (50.1) vanishes and the system (50) reduces to

(51.0) $x_1 = x_4 = 0,\quad x_2 = x_3 = x_5 = 1,$

(51.2) $9\bar{x}_8 + 2\bar{x}_6 + x_7 \geqslant 5,$

(51.3) $9\bar{x}_6 + 8x_7 \geqslant 8,$

(51.4) $3x_8 + \bar{x}_6 \geqslant 0;$

we have introduced the supplementary relation (51.0), in order to indicate the fixed variables which have led to the system (51).

In the above system, the single inequality in a determinate case is (51.4): as it is in case 1°B, it is redundant and so our system reduces to (51.2) and (51.3).

These inequalities are both in the "partially undeterminate" case 2°B; the p for (51.2) is 1, while for (51.3) it is 2; therefore we split the discussion into the three alternatives

$\alpha_1')$ $\bar{x}_6 = 1$;

$\alpha_2')$ $\bar{x}_6 = 0, \quad x_7 = 1$;

$\beta')$ $\bar{x}_6 = x_7 = 0$ and the remaining inequality would be the absurdity $0 \geqslant 8$; hence this alternative can be dropped.

In the case α_1', we obtain a single inequality to be solved:

(52.0) $$x_1 = x_4 = x_6 = 0, \quad x_2 = x_3 = x_5 = 1,$$

(52.2) $$9\bar{x}_8 + x_7 \geqslant 3.$$

According to the case 2^0B, we have two possibilities:

$\alpha'')$ $\bar{x}_8 = 1$,

$\beta'')$ $\bar{x}_8 = 0$ and $x_7 \geqslant 3$, which is inconsistent; hence this alternative can be dropped.

In the case α'' we have no more conditions, so that we have come to the solutions:

(53) $x_1=0,\ x_2=1,\ x_3=1,\ x_4=0,\ x_5=1,\ x_6=0,\ x_7$ arbitrary, $x_8=0$.

Now we come back to the alternative α_2', in which (51.3) is also verified; we have

(54.0) $$x_1 = x_4 = 0, \quad x_2 = x_3 = x_5 = x_6 = x_7 = 1,$$

(54.2) $$9\bar{x}_8 \geqslant 4,$$

hence $\bar{x}_8 = 1$ and we have obtained the following solution of the system (50) :

(55) $$x_1=0,\ x_2=1,\ x_3=1,\ x_4=0,\ x_5=1,\ x_6=1,\ x_7=1,\ x_8=0.$$

It remains the case β. We have

(56.0) $$x_3 = 0,$$

(56.1) $$6\bar{x}_5 + 4\bar{x}_2 + 3x_4 + 2x_1 = 8,$$

(56.2) $$9\bar{x}_8 + 5\bar{x}_1 + 3\bar{x}_5 + 2\bar{x}_6 + x_7 \geqslant 14,$$

(56.3) $$9\bar{x}_6 + 8x_7 + 6x_2 + 5\bar{x}_5 + 4x_1 + 4x_4 \geqslant 14,$$

(56.4) $$4\bar{x}_4 + 3x_8 + 2x_2 + \bar{x}_6 \geqslant 6.$$

The single relation in a "determinate" case is(56.2) which, according to the case 5°B, implies $\bar{x}_8 = 1$. Therefore:

(57.0) $$x_3 = x_8 = 0,$$

(57.1) $$6\bar{x}_5 + 4\bar{x}_2 + 3x_4 + 2x_1 = 8,$$

(57.2) $$5\bar{x}_1 + 3\bar{x}_5 + 2\bar{x}_6 + x_7 \geqslant 5,$$

(57.3) $$9\bar{x}_6 + 8x_7 + 6x_2 + 5\bar{x}_5 + 4x_1 + 4x_4 \geqslant 14,$$

(57.4) $$4\bar{x}_4 + 2x_5 + \bar{x}_6 \geqslant 6.$$

The single relation in a "determinate" case is(57.4), implying $\bar{x}_4 = 1$. Hence :

(58.0) $$x_3 = x_4 = x_8 = 0,$$

(58.1) $$6\bar{x}_5 + 4\bar{x}_2 + 2x_1 = 8,$$

(58.2) $$5\bar{x}_1 + 3\bar{x}_5 + 2\bar{x}_6 + \bar{x}_7 \geqslant 5,$$

(58.3) $$9\bar{x}_6 + 8x_7 + 6x_2 + 5\bar{x}_5 + 4x_1 \geqslant 14,$$

(58.4) $$2x_2 + \bar{x}_6 \geqslant 2.$$

Reasoning as above, the equation (58.1) implies $\bar{x}_5 = 1$, hence:

(59.0) $$x_3 = x_4 = x_5 = x_8 = 0,$$

(59.1) $$4\bar{x}_2 + 2x_1 = 2,$$

(59.2) $$5\bar{x}_1 + 2\bar{x}_6 + \bar{x}_7 \geqslant 2,$$

(59.3) $$9\bar{x}_6 + 8x_7 + 6x_2 + 4x_1 \geqslant 9,$$

(59.4) $$2x_2 + x_6 \geqslant 2.$$

By the same reasoning, the first equation (case 3^0 A) implies $\bar{x}_2 = 0$, hence :

(60.0) $$x_3 = x_4 = x_5 = x_8 = 0, \quad x_2 = 1,$$

(60.1) $$2x_1 = 2,$$

(60.2) $$5\bar{x}_1 + 2\bar{x}_6 + \bar{x}_7 \geqslant 2,$$

(60.3) $$9\bar{x}_6 + 8x_7 + 4x_1 \geqslant 3,$$

(60.4) $$\bar{x}_6 \geqslant 0.$$

The last inequality is identically satisfied (case 1^0B), while the equation (60.1) has the solution $x_1 = 1$. The system reduces to:

(61.0) $$x_3 = x_4 = x_5 = x_8 = 0, \; x_1 = x_2 = 1,$$

(61.2) $$2\bar{x}_6 + x_7 \geqslant 2,$$

(61.3) $$9\bar{x}_6 + 8x_7 \geqslant - 1.$$

The second inequality is identically satisfied (case 1^0B), while the first one leads to the cases

α"') $\bar{x}_6 = 1$,

β"') $\bar{x}_6 = 0$ and $\bar{x}_7 \geqslant 2$, which is inconsistent; hence this alternative can be dropped.

The alternative α"' corresponds to the solutions

(62) $x_1=1,\ x_2=1,\ x_3=0,\ x_4=0,\ x_5=0,\ x_6=0,\ x_7$ arbitrary, $x_8=0$.

We have thus found all the families of solutions of the system (50) :

- Table 12 -

x_1	x_2	x_3	x_4	x_5	x_6	x_7	x_8
0	1	1	0	1	0	-	0
0	1	1	0	1	1	1	0
1	1	0	0	0	0	-	0

Example 6. Let us solve the system

(63.1) $x_1 - 3x_2 + 12x_3 + x_5 - 7x_6 + x_7 - 3x_{10} + 5x_{11} + x_{12} - 6 \geqslant 0$,

(63.2) $-3x_1 + 7x_2 - x_4 - 6x_5 + 1 \geqslant 0$,

(63.3) $-11x_1 - x_3 + 7x_4 + x_6 - 2x_7 - x_8 + 5x_9 - 9x_{11} - 4 \geqslant 0$,

(63.4) $-5x_2 - 6x_3 + 12x_5 - 7x_6 - 3x_8 - x_9 + 8x_{10} - 5x_{12} + 8 \geqslant 0$,

(63.5) $7x_1 + x_2 + 5x_3 - 3x_4 - x_5 + 8x_6 + 2x_8 - 7x_9 - x_{10} + 7_{12} - 7 \geqslant 0$,

(63.6) $2x_1 + 4x_4 + 3x_7 + 5x_8 + x_9 - x_{11} - x_{12} - 4 \geqslant 0$,

which can be brought to the equivalent form

(64.1) $12x_3 + 7\bar{x}_6 + 5x_{11} + 3\bar{x}_2 + 3\bar{x}_{10} + x_1 + x_5 + x_7 + x_{12} \geqslant 19$,

(64.2) $7x_2 + 6\bar{x}_5 + 3\bar{x}_1 + x_4 \geqslant 9$,

(64.3) $11\bar{x}_1 + 9\bar{x}_{11} + 7x_4 + 5x_9 + 2\bar{x}_7 + \bar{x}_3 + x_6 + \bar{x}_8 \geqslant 28$,

(64.4) $12x_5 + 8x_{10} + 7\bar{x}_6 + 6\bar{x}_3 + 5\bar{x}_2 + 5\bar{x}_{12} + 3\bar{x}_8 + \bar{x}_9 \geqslant 19$,

(64.5) $8x_6 + 7x_1 + 7\bar{x}_9 + 7x_{12} + 5x_3 + 3\bar{x}_4 + 2x_8 + x_2 + \bar{x}_5 + \bar{x}_{10} \geqslant 19$,

(64.6) $5x_8 + 4x_4 + 3x_7 + 2x_1 + x_9 + \bar{x}_{11} + \bar{x}_{12} \geqslant 6$.

The inequality (64.3) is in the case 5°B, implying $\bar{x}_1 = 1$, that is $x_1 = 0$. We introduce this value in the system and we see that there is no inequality in a "determinate" case; relation (64.2) reduces to $7x_2 + 6\bar{x}_5 + \bar{x}_4 \geqslant 6$, which is in the case 2^oB. We have to consider the three alternatives:

α_1) $x_2 = 1$,

α_2) $x_2 = 0$, $\bar{x}_5 = 1$,

β) $x_2 = \bar{x}_5 = 0$ and $\bar{x}_4 \geqslant 6$, which is inconsistent; hence this alternative can be dropped.

We begin with the alternative α_1, which means that (64.2) is verified and $x_1 = 0$ (see above), $x_2 = 1$; these values reduce the inequality (64.1) to $12x_3 + 7\bar{x}_6 + 5x_{11} + 3\bar{x}_{1o} + x_5 + x_7 + x_{12} \geqslant 19$, which is in the case 5^oB, implying $x_3 = 1$. The values $x_1 = 0$, $x_2 = 1$, $x_3 = 1$ reduce (64.3) to the inequality $9\bar{x}_{11} + 7x_4 + 5x_9 + 2\bar{x}_7 + x_6 + \bar{x}_8 \geqslant 17$, which is again in case 5^oB, implying $\bar{x}_{11} = 1$. Now the inequality (64.1) is reduced to $7\bar{x}_6 + 3\bar{x}_{1o} + x_5 + x_7 + x_{12} \geqslant 7$, which, belonging to the case 5^oB, implies $\bar{x}_6 = 1$. The inequality (64.1) is transformed into an identity, so that the system (64) reduces to

(65.0) $\qquad x_1 = x_6 = x_{11} = 0, \; x_2 = x_3 = 1,$

(65.3) $\qquad 7x_4 + 5x_9 + 2\bar{x}_7 + \bar{x}_8 \geqslant 8,$

(65.4) $\qquad 12x_5 + 8x_{1o} + 5\bar{x}_{12} + 3\bar{x}_8 + \bar{x}_9 \geqslant 12,$

(65.5) $\qquad 7\bar{x}_9 + 7x_{12} + 3\bar{x}_4 + 2x_8 + \bar{x}_5 + \bar{x}_{1o} \geqslant 13,$

(65.6) $\qquad 5x_8 + 4x_4 + 3x_7 + x_9 + \bar{x}_{12} \geqslant 5.$

The inequalities (65.4) and (65.6) are in case 2^oB, while (65.3) and (65.5) belong to 6^oB. We perform the splitting resulting from (65.6):

α') $x_8 = 1$,

β') $x_8 = 0$ and $4x_4 + 3x_7 + x_9 + \bar{x}_{12} \geqslant 5$.

In the alternative α', the inequality (65.3) reduces to $7x_4 + 5x_9 + 2\bar{x}_7 \geqslant 8$, which is in the case 5^0B, implying $x_4 = 1$. Hence (65.5) becomes $7\bar{x}_9 + 7x_{12} + \bar{x}_5 + \bar{x}_{10} \geqslant 1$, therefore (case 5^0B) $\bar{x}_9 = 1$ and $7x_{12} + \bar{x}_5 + \bar{x}_{10} \geqslant 4$, implying $x_{12}=1$, again by the conclusion 5^0B. Now (65.3) becomes $2\bar{x}_7 \geqslant 1$, hence $\bar{x}_7 = 1$; further, (65.4) reduces to $12x_5 + 8x_{10} \geqslant 11$, implying $x_5 = 1$. These values satisfy the system (65), so that we have found the following solutions of (64) :

(66) $x_1=0,\ x_2=1,\ x_3=1,\ x_4=1,\ x_5=1,\ x_6=0,\ x_7=0,\ x_8=1,\ x_9=0,$

x_{10} arbitrary, $x_{11}=0,\ x_{12}=1$.

In the alternative β', the inequality (65.5) becomes $7\bar{x}_9 + 7x_{12} + 3\bar{x}_4 + \bar{x}_5 + \bar{x}_{10} \geqslant 13$, implying $\bar{x}_9 = 1$ and $7x_{12} + 3\bar{x}_4 + \bar{x}_5 + \bar{x}_{10} \geqslant 6$. We must take $x_{12} = 1$, otherwise we would obtain an inconsistent inequality. Further, (65.6) becomes $4x_4 + 3x_7 \geqslant 5$, implying $x_4 = 1$ and $3x_7 \geqslant 1$, hence $x_7 = 1$. Now the inequalities (65.3), (65.5) and (65.6) are verified, so that the system (65)reduces to (65.4), which becomes $12x_5 + 8x_{10} \geqslant 8$. This inequality is solved taking either $x_5 = 1$, or $x_5 = 0$ and $x_{10} = 1$, leading to the following solutions of the system (64):

(67) $x_1=0,\ x_2=1,\ x_3=1,\ x_4=1,\ x_5=1,\ x_6=0,\ x_7=1,\ x_8=0,$

$x_9=0,\ x_{10}$ arbitrary, $x_{11}=0,\ x_{12}=1,$

and

(68) $x_1=0,\ x_2=1,\ x_3=1,\ x_4=1,\ x_5=0,\ x_6=0,\ x_7=1,\ x_8=0$,

$x_9=0,\ x_{10}=1,\ x_{11}=0,\ x_{12}=1,$

respectively.

Now it remains the alternative α_2 :

(69.0) $x_1 = x_2 = x_5 = 0,$

(69.1) $12x_3 + 7\bar{x}_6 + 5x_{11} + 3\bar{x}_{1o} + x_7 + x_{12} \geqslant 16,$

(69.3) $9\bar{x}_{11} + 7x_4 + 5x_9 + 2\bar{x}_7 + \bar{x}_3 + x_6 + \bar{x}_8 \geqslant 17,$

(69.4) $8x_{1o} + 7\bar{x}_6 + 6\bar{x}_3 + 5\bar{x}_{12} + 3\bar{x}_8 + \bar{x}_9 \geqslant 14,$

(69.5) $8x_6 + 7\bar{x}_9 + 7x_{12} + 5x_3 + 3\bar{x}_4 + 2x_8 + \bar{x}_{1o} \geqslant 18,$

(69.6) $5x_8 + 4x_4 + 3x_7 + x_9 + \bar{x}_{11} + \bar{x}_{12} \geqslant 6.$

All these inequalities are in the case 6^oB. We shall split the discussion with respect to the variable x_8, i.e. : γ_1) $x_8 = 1$ and γ_2) $x_8 = 0$.

In the alternative γ_1, we have $x_8 = 1$ and (69.3) is reduced to $9\bar{x}_{11} + 7x_4 + 5x_9 + 2\bar{x}_7 + \bar{x}_3 + x_6 \geqslant 17$, which is in case 5^oB and implies $\bar{x}_{11} = 1$. Then (69.1) is transformed into $12x_3 + 7\bar{x}_6 + 3\bar{x}_{1o} + x_7 + x_{12} \geqslant 16$, implying $x_3 = 1$. Now the inequality (69.4) becomes $8x_{1o} + 7\bar{x}_6 + 5\bar{x}_{12} + \bar{x}_9 \geqslant 14$, implying thus $x_{1o} = 1$, hence (69.1) becomes $7\bar{x}_6 + x_7 + x_{12} \geqslant 4$. Since $\bar{x}_6 = 0$ would imply $x_7 + x_{12} \geqslant 4$, we must take $\bar{x}_6 = 1$; the inequality (69.1) is thus verified. Now (69.3) becomes $7x_4 + 5x_9 + 2\bar{x}_7 \geqslant 8$ and implies $x_4 = 1$; further, the inequality (69.5) is reduced to $7\bar{x}_9 + 7x_{12} \geqslant 11$, implying $\bar{x}_9 = 1$ and $7x_{12} \geqslant 4$, hence $x_{12} = 1$. Also, (69.1) reduces to $2\bar{x}_7 \geqslant 1$, i.e. to $\bar{x}_7 = 1$. The above found values satisfy the system (69), so that we have found the following solution of the system (64):

(70) $x_1=0,\ x_2=0,\ x_3=1,\ x_4=1,\ x_5=0,\ x_6=0,\ x_7=0,\ x_8=1,\ x_9=0,$

$$x_{1o}=1,\ x_{11}=0,\ x_{12}=1.$$

In the alternative γ_2, all the inequalities (69) are in the case 6^oB; we shall split the discussion with respect to x_{11} :

γ_1') $x_8 = 0$, $x_{11} = 1$, and γ_2') $x_8 = x_{11} = 0$.

In the alternative γ_1', the inequality (69.3) becomes $7x_4 + 5x_9 + 2\bar{x}_7 + \bar{x}_3 + x_6 \geqslant 16$, which implies $x_4 = x_9 = \bar{x}_7 = \bar{x}_3 = x_6 = 1$, so that (69.1) reduces to the inequality $3\bar{x}_{10} + x_{12} \geqslant 11$, which is inconsistent.

It remains the alternative γ_2', in which (69.1) reduces to $12x_3 + 7\bar{x}_6 + 3\bar{x}_{10} + x_7 + x_{12} \geqslant 16$ and implies $x_3 = 1$. Hence the system (69) becomes :

(71.0) $\quad x_1 = x_2 = x_5 = x_8 = x_{11} = 0, \quad x_3 = 1,$

(71.1) $\quad 7\bar{x}_6 + 3\bar{x}_{10} + x_7 + x_{12} \geqslant 4,$

(71.3) $\quad 7x_4 + 5x_9 + 2\bar{x}_7 + x_6 + \bar{x}_8 \geqslant 8,$

(71.4) $\quad 8x_{10} + 7\bar{x}_6 + 5\bar{x}_{12} + \bar{x}_9 \geqslant 11,$

(71.5) $\quad 8x_6 + 7\bar{x}_9 + 7x_{12} + 3\bar{x}_4 + \bar{x}_{10} \geqslant 13,$

(71.6) $\quad 4x_4 + 3x_7 + x_9 + \bar{x}_{11} + \bar{x}_{12} \geqslant 6.$

If $x_6 = 1$, then (71.1) becomes $3\bar{x}_{10} + x_7 + x_{12} \geqslant 4$, hence $\bar{x}_{10} = 1$, while (71.4) reduces to $8x_{10} + 5\bar{x}_{12} + \bar{x}_9 \geqslant 11$, implying $x_{10} = 1$, a contradiction.

If $x_6 = 0$, then the inequality (71.5) reduces to $7\bar{x}_9 + 7x_{12} + 3\bar{x}_4 + \bar{x}_{10} \geqslant 13$, implying $x_9 = 1$ and $7x_{12} + 3\bar{x}_4 + \bar{x}_{10} \geqslant 6$, hence $x_{12} = 1$. Now (71.3) becomes $7x_4 + 2\bar{x}_7 \geqslant 7$, hence $x_4=1$, while (71.4) reduces to $8x_{10} \geqslant 3$, implying thus $x_{10}=1$. Further, (71.6) is transformed into $3x_7 \geqslant 1$, i.e., $x_7 = 1$. These values satisfy the system (71) so that we have found the last solution of the system (64) :

(72) $x_1=0,\ x_2=0,\ x_3=1,\ x_4=1,\ x_5=0,\ x_6=0,\ x_7=1,\ x_8=0,$

$$x_9=0,\ x_{10}=1,\ x_{11}=0,\ x_{12}=1.$$

Thus the table of all the solutions of (64) is the following :

- Table 13 -

x_1	x_2	x_3	x_4	x_5	x_6	x_7	x_8	x_9	x_{10}	x_{11}	x_{12}
0	1	1	1	1	0	0	1	0	-	0	1
0	1	1	1	1	0	1	0	0	-	0	1
0	1	1	1	0	0	1	0	0	1	0	1
0	0	1	1	0	0	0	1	0	1	0	1
0	0	1	1	0	0	1	0	0	1	0	1

§ 5. Computational Status

The above procedures were tested on several problems solved by hand computation, and the results seem to be satisfactory. So, for instance, Example 5 of § 3, having 6 inequalities with 12 unknowns, was solved in less then one hundred minutes, (we recall that by direct inspection we should be faced with the checking of $2^{12} = 4096$ variants).

The programming of the method for an ELLIOTT-8o3 B and for a MECIPT-1 computers are in progress.

References

1. E.BALAS : An Additive Algorithm for Solving Linear Programs with Zero-One Variables. Operations Research,13,517-546 (1965).

2. R.BELLMAN : Combinatorial Processes and Dynamic Programming. Combinatorial Analysis, Proceedings of Symposia in Applied Mathematics, Vol.X, 1960, pp.217-250.

3. P.BERTIER-PH.T.NGHIEM : Résolution de problèmes en variables bivalents (Algorithme de Balas et procèdure SEP). SEMA, Note de travail No.33, Janvier 1965.

4. P.BERTIER-PH.T.NGHIEM, B.ROY: Procédure SEP. Trois exemples numériques. SEMA, Note de travail No. 32, Janvier 1965.

5. P.BERTIER-B.ROY : Une procédure de résolution pour une classe de problèmes pouvant avoir un caractère combinatoire. SEMA, Note de travail No.30 bis, Décembre 1964.

6. P.CAMION : Quelques propriétés des chemins et circuits hamiltoniens dans la théorie des graphes. Cahiers du Centre d'Etudes de Recherche Opérationnelle, Bruxelles , vol.2, No.1, 5-36, 1960.

7. P.CAMION : Une méthode de résolution par l'algèbre de Boole des problèmes combinatoires où interviennent des entiers. Cahiers du Centre d'Etudes de Recherche Opérationnelle, 2, 234-289, 1960.

8. M.CARVALLO : Monographie des treillis et algèbre de Boole. Paris, Gauthier-Villars, 1962.

9. M.CARVALLO : Principes et applications de l'analyse boolénne. Paris, Gauthier-Villars, 1965.

10. G.B.DANTZIG : Discrete Variable Extremum Problems. Operations Research, 5, No.2, 266-277 (1957).

11. G.B.DANTZIG : On the Significance of Solving Linear Programming Problems with Some Integer Variables. Econometrica, 28, No.1, 30-44 (1960).

12. G.B.DANTZIG : Linear Programming and Extensions. Ch. 26. Princeton Univ. Press, Princeton, 1963.

13. M.DENIS-PAPIN - R.FAURE - A.KAUFMANN : Cours de calcul booléien. Editions Albin-Michel, Paris 1963.

14. R.FAURE - Y.MALGRANGE : Nouvelles recherches sur la résolution des programmes linéaires en nombres entiers. Gestion, No.spécial, Juin 1965, pp.371-375.

15. R.FORTET : L'algèbre de Boole et ses applications en recherche opérationnelle. Cahiers du Centre d'Etudes de Recherche Opérationnelle, Bruxelles, vol.1,n.4,5-36, 1959.

16. R.FORTET : Application de l'algèbre de Boole en recherche opérationnelle. Revue Française de Recherche Opérationnelle, Paris, vol.4, no.14, 17-26, 1960.

17. R.FORTET : Résolution booléenne d'opérations arithmétiques sur les entiers non négatifs et application aux programmes linéaires en nombres entiers, SEMA, Paris, Mars 1960.

18. R.E.GOMORY : Essentials of an Algorithm for Integer Solutions to Linear Programs. Bull.Amer. Math. Soc., 64, No.5, 275-278 (1958).

19. R.E.GOMORY : An Algorithm for Integer Solutions to Linear Programs. Princeton-IBM Math. Research Project,Techn. Report No.1, November 17,1958.Republished in Recent

Advances in Mathematical Programming, edited by R.L. Graves and Ph.Wolfe, McGraw Hill, New York, 1963.

20. P.L.IVANESCU : Systems of Pseudo-Boolean Equations and Inequalities. Bulletin de l'Académie Polonaise des Sciences, 12, n.11, 673-680 (1964).

21. P.L.IVANESCU : The Method of Successive Eliminations for Pseudo-Boolean Equations. Bulletin de l'Académie Polonaise des Sciences, 12, n.11, 681-683 (1964).

22. P.L.IVANESCU : Pseudo-Boolean Programming and Applications (Abstract of Doctor's Thesis), Lecture Notes in Mathematics, No.9, 1965, Springer Verlag, Berlin-Heidelberg-New York.

23. P.L.IVANESCU : Pseudo-Boolean Programming with Special Restraints. Applications to Graph Theory. Elektronische Informationsverarbeitung und Kybernetik (E.I.K.), 1, No.3, 167-185 (1965).

24. P.L.IVANESCU : Dynamic Programming with Bivalent Variables. Lecture at the Symposium on Applications of Mathematics to Economics, Smolenice (Czechoslovakia), June 1965. To appear in Publ. Inst. Math. Belgrade.

25. P.L.IVANESCU, I.ROSENBERG, S.RUDEANU : On the Determination of the Minima of Pseudo-Boolean Functions (in Romanian). Studii şi Cercetări Matematice, 14,No.3, 359-364 (1963).

26. P.L.IVANESCU, I.ROSENBERG, S.RUDEANU : An Application of Discrete Linear Programming to the Minimization of Boolean Functions (in Russian). Revue Math. Pures et Appl., 8, No.3, 459-475 (1963).

27. A.KAUFMANN : Méthodes et modèles de la recherche opérationnelle. Tome 2, Dunod, Paris, 1964.

28. A.KAUFMANN - Y.MALGRANGE : Recherche des chemins et circuits hamiltoniens d'un graphe. Revue Française de Recherche Opérationnelle, 7, 61-73 (1963).

29. K.MAGHOUT : Applications de l'algèbre de Boole à la théorie des graphes et aux programmes linéaires et quadratiques. Cahiers du Centre d'Etudes de Recherche Opérationnelle, Bruxelles, vol.5, n.1-2, 21-99 (1963).

30. B.ROY : Cheminement et connexité dans les graphes. Application aux problèmes d'ordonnancement. METRA, Série Spéciale, n.1, 1962.

31. B.ROY, PH.T.NGHIEM, P.BERTIER : Programmes linéaires en nombres entiers et procédure SEP. METRA, 4,No.3,441-460 (1963).

32. B.ROY - B.SUSSMANN : Les problèmes d'ordonnancement avec contraintes disjonctives. SEMA, Rapport de Recherche No.9 bis, Octobre 1964.

33. S.RUDEANU : Irredundant Solutions of Boolean and Pseudo-Boolean Equations. Rev.Roumaine Math.Pures et Appl., 11, 183-188, 1966.

Part II

NONLINEAR PSEUDO-BOOLEAN EQUATIONS AND INEQUALITIES

In Part I we have proposed a method for the determination of all the solutions of a system of linear pseudo-Boolean equations and/or inequalities. The aim of Part II is to solve the problem in case of a system of arbitrary (i.e. linear and/or nonlinear), equations and/or inequalities.

We recall that a Boolean function has bivalent (0, 1) variables and bivalent values, while a pseudo-Boolean function has again bivalent values, but takes real values.

In Part II we associate to each pseudo-Boolean equation (or inequality, or system of equations and/or inequalities) a "characteristic" Boolean equation which has the same solutions as the original system (§§ 1, 2, 3). This idea allows also the inclusion of logical conditions in the system (§ 4).

The construction of the characteristic equation is based on the reduction of the general case to the linear one*) ; this "linearization" process does not raise computational difficulties.

---

*) Another "linearization" process was proposed by R.FORTET /4/.

The problem is now reduced to that of solving the characteristic equation. This task is done using a procedure which gives the solutions grouped into pairwise disjoint " families of solutions" (§§ 5,6).

We suppose that the reader is familiar with the elements of Boolean calculus and with Part I.

§ 1. The Characteristic Function in the Linear Case

Let $\sum(x_1,\ldots,x_n)$ denote a pseudo-Boolean equation, or inequality, or system of pseudo-Boolean equations and/or inequalities.

Definition 1. The characteristic equation of $\sum(x_1,\ldots,x_n)$ is a Boolean equation

(1) $$\Phi(x_1,\ldots,x_n) = 1$$

which has the same solutions as $\sum(x_1,\ldots,x_n)$; the Boolean function $\Phi(x_1,\ldots,x_n)$ will be called the <u>characteristic function</u>*) of $\sum(x_1,\ldots,x_n)$.

In other words, the characteristic function of a pseudo-Boolean system is simply the characteristic function of the set of its solutions.

Now, we recall the well-known interpolation formula for Boolean functions:

(2) $$\Psi(x_1,\ldots,x_n) = \bigcup_{\alpha_1,\ldots,\alpha_n} \Psi(\alpha_1,\ldots,\alpha_n)\, x_1^{\alpha_1} \ldots x_n^{\alpha_n}$$

---

*) In /7/ and /8/ this function was termed the "reduct"of $\sum$.

where $\bigcup_{\alpha_1,\ldots,\alpha_n}$ means that the disjunction is extended over all 2^n possible systems of values 0,1 of $\alpha_1,\ldots,\alpha_n$, and the notation x^{α} means

(3) $$x^{\alpha} = \begin{cases} x, & \text{if } \alpha = 1, \\ \bar{x}, & \text{if } \alpha = 0. \end{cases}$$

In other words, we have

(4) $$\Psi(x_1,\ldots,x_n) = \bigcup_{\alpha_1,\ldots,\alpha_n}^{1} x_1^{\alpha_1} \ldots x_n^{\alpha_n}$$

where $\bigcup_{\alpha_1,\ldots,\alpha_n}^{1}$ means that the disjunction is extended only over those values of the vector $(\alpha_1,\ldots,\alpha_n)$ for which $\Psi(\alpha_1,\ldots,\alpha_n) = 1$.

Therefore, the characteristic function Φ of $\sum(x_1,\ldots,x_n)$ is given by the following formula:

(5) $$\Phi(x_1,\ldots,x_n) = \bigcup_{\alpha_1,\ldots,\alpha_n}^{\Sigma} x_1^{\alpha_1} \ldots x_n^{\alpha_n},$$

where $\bigcup_{\alpha_1,\ldots,\alpha_n}^{\Sigma}$ means that the disjunction is extended over all the solutions $(\alpha_1,\ldots,\alpha_n)$ of $\sum(x_1,\ldots,x_n)$.

Now, the results of Part I allow the immediate construction of the characteristic function in the linear case. The necessity of this construction will become clear in the next section.

a) <u>Linear Equations</u>

In the case of a single linear pseudo-Boolean equation, the knowledge of all the solutions (obtained, for instance, as in Part I), permits the direct determination of the characteristic function, via formula (5).

Example 1. The linear equation (21) in Example 2 of Part I, § 2, i.e.

$$(6)\quad 4x_1 + \bar{x}_1 - 3x_2 + \bar{x}_2 + 5x_3 - 2x_4 + 5x_5 + 2x_6 - x_7 = 7$$

was shown to have the solutions

- Table 1 -

x_1	x_2	x_3	x_4	x_5	x_6	x_7
0	1	1	0	1	0	1
0	1	1	1	1	1	1
1	0	1	1	0	0	1
0	0	1	0	0	0	0
0	0	1	1	0	1	0
1	1	1	0	0	1	1
1	0	0	1	1	0	1
0	0	0	0	1	0	0
0	0	0	1	1	1	0
1	1	0	0	1	1	1
1	0	0	0	0	1	0

Hence the characteristic function is

$$(7)\qquad \Phi_1 = \bar{x}_1x_2x_3\bar{x}_4x_5\bar{x}_6x_7 \cup \bar{x}_1x_2x_3x_4x_5x_6x_7 \cup$$
$$\cup x_1\bar{x}_2x_3x_4\bar{x}_5\bar{x}_6x_7 \cup \bar{x}_1\bar{x}_2x_3\bar{x}_4\bar{x}_5\bar{x}_6\bar{x}_7 \cup$$
$$\cup \bar{x}_1\bar{x}_2x_3x_4\bar{x}_5x_6\bar{x}_7 \cup x_1x_2x_3\bar{x}_4\bar{x}_5x_6x_7 \cup$$
$$\cup x_1\bar{x}_2\bar{x}_3x_4x_5\bar{x}_6x_7 \cup \bar{x}_1\bar{x}_2\bar{x}_3\bar{x}_4x_5\bar{x}_6\bar{x}_7 \cup$$
$$\cup \bar{x}_1\bar{x}_2\bar{x}_3x_4x_5x_6\bar{x}_7 \cup x_1x_2\bar{x}_3\bar{x}_4x_5x_6x_7 \cup$$
$$\cup x_1\bar{x}_2\bar{x}_3\bar{x}_4\bar{x}_5x_6\bar{x}_7 .$$

b) Linear Inequalities

In Part I, the solutions of a linear inequality were given grouped into "families of solutions". A family $\mathcal{F}$ of solutions was defined as being a set of solutions characterized by the fact that certain variables have fixed values, while the other remain arbitrary:

$$(8) \quad \mathcal{F}: x_{h_1} = \xi_{h_1}, \ldots, x_{h_m} = \xi_{h_m}; \; x_{h_{m+k}} \text{ arbitrary for } k=1,\ldots,n-m.$$

In Part I we have indicated a procedure for obtaining a system $\mathcal{F}_1, \ldots, \mathcal{F}_p$ of families of solutions with the property that the set $\mathcal{S}$ of all the solutions is expressed as the set-theoretical join

$$(9) \qquad \mathcal{S} = \mathcal{F}_1 \cup \ldots \cup \mathcal{F}_p.$$

If we take into account relation (9) and the idempotency law ($z = z \cup z = z \cup z \cup z =$ etc.), formula (5) becomes

$$(10) \quad \Phi(x_1, \ldots, x_n) = \bigcup_{h=1}^{p} \bigcup_{(\alpha_1, \ldots, \alpha_n) \in \mathcal{F}_h} x_1^{\alpha_1} \ldots x_n^{\alpha_n}.$$

Let us now notice that for a family $\mathcal{F}$, formula (8) implies

$$(11) \qquad \bigcup_{(\alpha_1, \ldots, \alpha_n) \in \mathcal{F}} x_1^{\alpha_1} \ldots x_n^{\alpha_n} = x_{h_1}^{\xi_{h_1}} \ldots x_{h_m}^{\xi_{h_m}}$$

because the left-hand side of (11) is, in fact, equal to

$$x_{h_1}^{\xi_{h_1}} \ldots x_{h_m}^{\xi_{h_m}} \bigcup_{\alpha_{h_{m+1}}, \ldots, \alpha_{h_n}} x_{h_{m+1}}^{\alpha_{h_{m+1}}} \ldots x_{h_n}^{\alpha_{h_n}}$$

where, $\alpha_{h_{m+1}}, \ldots, \alpha_{h_n}$ taking all possible values 0 and 1, make the last disjunction equal to 1.

Now, applying formula (11), we see that each family $\mathcal{F}_h (h=1,\ldots,p)$ is characterized by a conjunction

(12) $$x_{h_1}^{\xi_{h_1}} \ldots x_{h_{m(h)}}^{\xi_{h_{m(h)}}} = C_h.$$

Therefore, from (10), (11) and (12) we have

THEOREM 1. <u>The characteristic function of a linear inequality may be written as</u>

(13) $$\Phi(x_1,\ldots,x_n) = C_1 \cup \ldots \cup C_p.$$

Example 2. The linear inequality (35) in Example 3 of Part I, § 3, i.e.

(14) $$2\bar{x}_1 - 5x_2 + 3x_3 + 4\bar{x}_4 - 7x_5 + 16x_6 - x_7 \geqslant - 4,$$

was shown to have the following families of solutions:

- Table 2 -

No.	x_1	x_2	x_3	x_4	x_5	x_6	x_7
1	-	-	-	-	-	1	-
2	-	0	-	-	0	0	-
3	-	1	-	0	0	0	-
4	-	1	1	1	0	0	-
5	0	1	0	1	0	0	-
6	-	0	-	0	1	0	-
7	0	0	1	1	1	0	-
8	1	0	1	1	1	0	0
9	0	1	1	0	1	0	-

This table indicates, for each family, the values of the fixed variables (0 or 1) and the variables which are arbitrary (dashes).

Formula (13) shows that the corresponding characteristic function is

$$(14)\quad \Phi_2 = x_6 \cup \bar{x}_2\bar{x}_5\bar{x}_6 \cup x_2\bar{x}_4\bar{x}_5\bar{x}_6 \cup x_2x_3x_4\bar{x}_5\bar{x}_6 \cup \bar{x}_1x_2\bar{x}_3x_4\bar{x}_5\bar{x}_6 \cup$$
$$\cup \bar{x}_2\bar{x}_4x_5\bar{x}_6 \cup \bar{x}_1\bar{x}_2x_3x_4x_5\bar{x}_6 \cup x_1\bar{x}_2x_3x_4x_5\bar{x}_6\bar{x}_7 \cup$$
$$\cup \bar{x}_1x_2x_3\bar{x}_4x_5\bar{x}_6 .$$

c) <u>Linear Systems</u>

In Part I the solutions of a system of linear pseudo-Boolean equations and/or inequalities were also grouped into families of solutions. Therefore the characteristic function of a linear system may be obtained in the same way as in the case of a single linear inequality.

Example 3. The linear system (49) in Example 5 of Part I, § 4, i.e.

$$(15.1)\qquad 2x_1 - 4x_2 + 8x_3 + 3x_4 - 6x_5 = -2,$$

$$(15.2)\qquad 5x_1 - 4x_3 + 3x_5 + 2x_6 - x_7 + 9x_8 \leq 5,$$

$$(15.3)\qquad 4x_1 + 6x_2 + 4x_4 - 5x_5 - 9x_6 + 8x_7 > -1,$$

$$(15.4)\qquad 2x_2 - 4x_4 - x_6 + 3x_8 \geq 1,$$

was shown to have the following solutions :

- Table 3 -

x_1	x_2	x_3	x_4	x_5	x_6	x_7	x_8
0	1	1	0	1	0	-	0
0	1	1	0	1	1	1	0
1	1	0	0	0	0	-	0

Hence the characteristic function is

(16) $$\Phi_3 = \bar{x}_1x_2x_3\bar{x}_4x_5\bar{x}_6\bar{x}_8 \cup \bar{x}_1x_2x_3\bar{x}_4x_5x_6x_7\bar{x}_8 \cup$$
$$\cup\, x_1x_2\bar{x}_3\bar{x}_4\bar{x}_5\bar{x}_6\bar{x}_8.$$

§ 2. The Characteristic Function for a Nonlinear Equation or Inequality

Let us consider a nonlinear pseudo-Boolean equation with the unknowns $x_1,\dots,x_n$:

(17) $$a_1P_1 + \dots + a_mP_m = b,$$

where each P_i $(i=1,\dots,m)$ stands for a certain conjunction (i.e. a product of variables with or without bars):

(18) $$P_i = x_{i_1}^{\pi_{i_1}} \dots x_{i_{k(i)}}^{\pi_{i_{k(i)}}}$$

Let us replace the product P_i by a single bivalent variable y_i and solve the resulting linear pseudo-Boolean equation

(19) $$a_1y_1 + \dots + a_my_m = b,$$

where $y_1,\dots,y_m$ are treated as independent variables.

If $\Psi(y_1,\dots,y_m)$ is the characteristic equation of (19), obtained as in § 1 (a), then the Boolean function

(20) $$\Phi(x_1,\dots,x_n) = \Psi\left(x_{1_1}^{\pi_{1_1}} \dots x_{1_{k(1)}}^{\pi_{1_{k(1)}}},\dots,x_{m_1}^{\pi_{m_1}} \dots x_{m_{k(m)}}^{\pi_{m_{k(m)}}}\right)$$

will be the characteristic function of (17).

In the case of a linear inequality we apply the same procedure.

Example 4. Let us solve the pseudo-Boolean equation

(21) $$- 6x_1\bar{x}_2x_3 - 4x_2x_4 + 2x_2x_4\bar{x}_5 + 4\bar{x}_3\bar{x}_4 = - 2$$

Putting

(22) $$x_1\bar{x}_2x_3 = y_1, \; x_2x_4 = y_2, \; x_2x_4\bar{x}_5 = y_3, \; \bar{x}_3\bar{x}_4 = y_4,$$

we have the linear equation

(23) $$- 6y_1 - 4y_2 + 2y_3 + 4y_4 = - 2,$$

which may be solved as in Part I and has the solutions

(24.1) $$y_1 = 0, \; y_2 = 1, \; y_3 = 1, \; y_4 = 0,$$

and

(24.2) $$y_1 = 1, \; y_2 = 0, \; y_3 = 0, \; y_4 = 1.$$

Hence the characteristic function of (23) is

(25) $$\varphi_1 = \bar{y}_1y_2y_3\bar{y}_4 \cup y_1\bar{y}_2\bar{y}_3y_4 \;;$$

from (20) and (22) we derive the characteristic function of (21):

$$\Phi_4 = (\bar{x}_1 \cup x_2 \cup x_3).x_2x_4.x_2x_4\bar{x}_5.(x_3 \cup x_4) \cup$$
$$\cup x_1\bar{x}_2x_3(\bar{x}_2 \cup \bar{x}_4)(\bar{x}_2 \cup \bar{x}_4 \cup x_5)\; \bar{x}_3\bar{x}_4,$$

or else

(26) $$\Phi_4 = x_2x_4\bar{x}_5.$$

The characteristic equation $\Phi_4 = 1$ shows that the solutions of (21) are:

(27) $$x_2 = x_4 = 1, \; x_5 = 0, \; x_1 \text{ and } x_3 \text{ arbitrary.}$$

Example 5. In order to solve the pseudo-Boolean inequality

(28) $$7x_1x_2x_3 + 5x_2x_4x_6x_7x_8 - 4x_3x_8 - 2\bar{x}_1x_4x_8 - x_4\bar{x}_5x_6 \leqslant 3,$$

we set

(29) $$x_1x_2x_3 = y_1,\ x_2x_4x_6x_7x_8 = y_2,\ x_3x_8 = y_3,\ \bar{x}_1x_4x_8 = y_4,$$
$$x_4\bar{x}_5x_6 = y_5.$$

Thus we obtain the inequality

(30) $$7y_1 + 5y_2 - 4y_3 - 2y_4 - y_5 \leqslant 3,$$

whose families of solutions, obtained as in Part I, are

- Table 4 -

y_1	y_2	y_3	y_4	y_5
0	0	-	-	-
0	1	1	-	-
0	1	0	1	-
1	0	1	-	-

leading to the characteristic function

(31) $$\Psi_2 = \bar{y}_1\bar{y}_2 \cup \bar{y}_1y_2y_3 \cup \bar{y}_1y_2\bar{y}_3y_4 \cup y_1\bar{y}_2y_3 =$$
$$= \bar{y}_1(\bar{y}_2 \cup y_3 \cup y_4) \cup \bar{y}_2y_3,$$

hence the characteristic function of (21) is

(32) $$\Phi_5 = (\bar{x}_1 \cup \bar{x}_2 \cup \bar{x}_3)(\bar{x}_2 \cup \bar{x}_4 \cup \bar{x}_6 \cup \bar{x}_7 \cup \bar{x}_8 \cup x_3 \cup \bar{x}_1)$$
$$\cup x_3x_8(\bar{x}_2 \cup \bar{x}_4 \cup \bar{x}_6 \cup \bar{x}_7) =$$
$$\bar{x}_1 \cup \bar{x}_2 \cup \bar{x}_3\bar{x}_4 \cup \bar{x}_3\bar{x}_6 \cup \bar{x}_3\bar{x}_7 \cup \bar{x}_3\bar{x}_8 \cup \bar{x}_4x_8 \cup \bar{x}_6x_8 \cup \bar{x}_7x_8 .$$

If we are interested*) in simultaneously obtaining the solutions of the equation with integer coefficients

$$f(x_1,\dots,x_n) = 0,$$

of the corresponding inequality

---

*) As it will be the case in Part III.

$$f(x_1,\ldots,x_n)\geqslant 0$$

and of the strict inequality

$$f(x_1,\ldots,x_n) > 0,$$

we "linearize" them as above and proceed as in Theorem 7 of Part I.

Example 6. Let us consider the equation

(35) $-7x_1x_2x_3 - 5x_2x_4x_6x_7x_8 + 4x_3x_8 + 2\bar{x}_1x_4x_8 + x_1\bar{x}_5x_6 = -3$

and the inequalities

(33) $-7x_1x_2x_3 - 5x_2x_4x_6x_7x_8 + 4x_3x_8 + 2\bar{x}_1x_4x_8 + x_1\bar{x}_5x_6 \geqslant -3,$

(34) $-7x_1x_2x_3 - 5x_2x_4x_6x_7x_8 + 4x_3x_8 + 2\bar{x}_1x_4x_8 + x_1\bar{x}_5x_6 > -3.$

With the substitutions (29), we obtain

(35) $$-7y_1 - 5y_2 + 4y_3 + 2y_4 + y_5 = -3$$

instead of (33),

(36) $$-7y_1 - 5y_2 + 4y_3 + 2y_4 + y_5 \geqslant -3$$

instead of (28), and

(37) $$-7y_1 - 5y_2 + 4y_3 + 2y_4 + y_5 > -3$$

instead of (34).

As in Part I, we seek the basic solutions of the canonical form of (36), i.e. of

(38) $$7\bar{y}_1 + 5\bar{y}_2 + 4y_3 + 2y_4 + y_5 \geqslant 9$$

and obtain

- Table 5 -

y_1	y_2	y_3	y_4	y_5	(35)?
1	1	0	0	0	
1	0	1	0	0	
1	0	0	1	0	✓
0	1	1	0	0	✓

We see that the equation (35) has the solutions

(39.1) $y_1 = 0,\ y_2 = 1,\ y_3 = 0,\ y_4 = 1,\ y_5 = 0$

and

(39.2) $y_1 = 1,\ y_2 = 0,\ y_3 = 1,\ y_4 = 0,\ y_5 = 0$

Hence, the characteristic function of (35) is

(40) $$\Psi_3 = \bar{y}_1 y_2 \bar{y}_3 y_4 \bar{y}_5 \cup y_1 \bar{y}_2 y_3 \bar{y}_4 \bar{y}_5 .$$

As in Part I we obtain now the basic solutions of the canonical form of the strict inequality (37):

- Table 6 -

$\bar{y}_1$	$\bar{y}_2$	y_3	y_4	y_5
1	1	0	0	0
1	0	1	0	0
1	0	0	1	1
0	1	1	1	0
0	1	1	0	1

Hence, the families of solutions of (37) are:

- Table 7 -

y_1	y_2	y_3	y_4	y_5
0	0	-	-	-
0	1	1	-	-
0	1	0	1	1
1	0	1	1	-
1	0	1	0	1

and its characteristic function is

(41) $\Psi_4 = \bar{y}_1\bar{y}_2 \cup \bar{y}_1 y_2 y_3 \cup \bar{y}_1 y_2 \bar{y}_3 y_4 y_5 \cup y_1 \bar{y}_2 y_3 y_4 \cup y_1 \bar{y}_2 y_3 \bar{y}_4 y_5 =$

$$= \bar{y}_1\bar{y}_2 \cup \bar{y}_1 y_3 \cup \bar{y}_1 y_4 y_5 \cup \bar{y}_2 y_3 y_4 \cup \bar{y}_2 y_3 y_5 .$$

Now, the characteristic function of (28) was obtained in Example 5. From (29) and (40) we deduce the characteristic function of (33):

(42) $$\Phi_6 = \bar{x}_1 x_2 \bar{x}_3 x_4 x_5 x_6 x_7 x_8 \cup x_1 x_2 x_3 \bar{x}_4 x_8 \cup \cup x_1 x_2 x_3 \bar{x}_6 x_8 \cup x_1 x_2 x_3 x_5 \bar{x}_7 x_8 ,$$

while from (29) and (41) we obtain the characteristic function of (34):

(43) $$\Phi_7 = \bar{x}_2 \cup (\bar{x}_1 \cup \bar{x}_3)(\bar{x}_4 \cup \bar{x}_6 \cup \bar{x}_7 \cup \bar{x}_8) \cup \bar{x}_1 (x_3 \cup \bar{x}_5) \cup \cup x_3 x_4 \bar{x}_5 x_6 \bar{x}_7 x_8 .$$

§ 3. The Characteristic Function for Systems

Let us consider a system of pseudo-Boolean equations and inequalities:

(44.j) $$f_j(x_1,\ldots,x_n) = 0 \quad (j=1,\ldots,m)$$

(44.h) $$f_h(x_1,\ldots,x_n) \geqslant 0 \quad (h=m+1,\ldots,m+p)$$

which can contain in case that the coefficients are integers, also inequalities of the form

(44.k) $$f_k(x_1,\ldots,x_n) > 0 \quad (k=m+p+1,\ldots,q)$$

and let

(45.1) $$\varphi_1(x_1,\ldots,x_n) = 1$$

$$\cdots\cdots\cdots$$

(45.9) $$\varphi_q(x_1,\ldots,x_n) = 1$$

be the corresponding characteristic equations, determined as in §§ 1, 2. If we denote by Φ the characteristic function of the system (44), we have obviously:

THEOREM 2.

$$\Phi(x_1,\ldots,x_n) = \prod_{s=1}^{q} \varphi_s(x_1,\ldots,x_n) \tag{46}$$

Example 7. Let us consider the system

$$7x_1x_2x_3 - 2\bar{x}_1x_4x_8 + 5x_2x_4x_6x_7x_8 - 4x_3x_8 - x_4\bar{x}_5x_6 \leq 3, \tag{47.1}$$

$$3x_1 - 2x_2\bar{x}_6 + 14x_5\bar{x}_6\bar{x}_8 + 2x_1x_2x_3 - 7x_8 \geqslant -8, \tag{47.2}$$

$$8x_4x_5\bar{x}_8 - 4\bar{x}_3\bar{x}_7x_8 + 3x_1x_2 + \bar{x}_3 + \bar{x}_4 + \bar{x}_5 < 3, \tag{47.3}$$

$$2x_3 + 3x_5 - \bar{x}_5x_6 + 4x_6\bar{x}_7x_8 - 2x_5x_6x_7x_8 \geqslant 1 . \tag{47.4}$$

The characteristic functions of the above inequalities, obtained as in § 2, are :

$$\varphi_1 = \bar{x}_1 \cup \bar{x}_2 \cup \bar{x}_3\bar{x}_8 \cup (\bar{x}_3 \cup x_8)(\bar{x}_4 \cup \bar{x}_6 \cup \bar{x}_7), \tag{48.1}$$

$$\varphi_2 = x_1 \cup \bar{x}_2 \cup x_6 \cup \bar{x}_8, \tag{48.2}$$

$$\varphi_3 = \bar{x}_3\bar{x}_7x_8 \cup (\bar{x}_1 \cup \bar{x}_2)(x_3 \cup x_4 \cup x_5)(\bar{x}_4 \cup \bar{x}_5 \cup x_8), \tag{48.3}$$

$$\varphi_4 = x_3 \cup x_5 \cup x_6\bar{x}_7x_8, \tag{48.4}$$

or else

$$\varphi_1 = \bar{x}_1 \cup \bar{x}_2 \cup \bar{x}_3\bar{x}_4 \cup \bar{x}_3\bar{x}_6 \cup \bar{x}_3\bar{x}_7 \cup \bar{x}_3\bar{x}_8 \cup \bar{x}_4x_8 \cup \bar{x}_6x_8 \cup \bar{x}_7x_8, \tag{48'.1}$$

$$\varphi_2 = x_1 \cup \bar{x}_2 \cup x_6 \cup \bar{x}_8, \tag{48'.2}$$

$$\begin{aligned}\varphi_3 = {} & \bar{x}_1x_3\bar{x}_4 \cup \bar{x}_1x_3\bar{x}_5 \cup \bar{x}_1x_3x_8 \cup \bar{x}_1x_4\bar{x}_5 \cup \bar{x}_1x_4x_8 \cup \\ & \cup \bar{x}_1\bar{x}_4x_5 \cup \bar{x}_1x_5x_8 \cup \bar{x}_2x_3\bar{x}_4 \cup \bar{x}_2x_3\bar{x}_5 \cup \bar{x}_2x_3x_8 \cup \\ & \cup \bar{x}_2x_4\bar{x}_5 \cup \bar{x}_2\bar{x}_4x_8 \cup \bar{x}_2\bar{x}_4x_5 \cup \bar{x}_2x_5x_8 \cup \bar{x}_3x_7x_8 ,\end{aligned} \tag{48'.3}$$

$$\varphi_4 = x_3 \cup x_5 \cup x_6\bar{x}_7x_8. \tag{48'.4}$$

Multiplying these functions, as indicated in Theorem 2, we obtain the characteristic function of the system:

$$(49)\quad \Phi_8 = x_1\bar{x}_3x_5\bar{x}_7x_8 \cup \bar{x}_3x_6\bar{x}_7x_8 \cup \bar{x}_2x_3\bar{x}_4 \cup \bar{x}_2x_3\bar{x}_5 \cup$$
$$\cup \bar{x}_2x_3x_8 \cup \bar{x}_2\bar{x}_4x_5 \cup \bar{x}_2x_5x_8 \cup \bar{x}_2x_6\bar{x}_7x_8 \cup \bar{x}_1x_3\bar{x}_5x_6 \cup$$
$$\cup \bar{x}_1x_3x_6x_8 \cup \bar{x}_1x_5x_6x_8 \cup \bar{x}_1x_6\bar{x}_7x_8 \cup \bar{x}_1x_3\bar{x}_5\bar{x}_8 \cup \bar{x}_1\bar{x}_4x_5\bar{x}_8.$$

From Theorem 2 we deduce:

Corollary 1. If the conditions in the original system are grouped into several subsystems $\Sigma_1,\ldots,\Sigma_r$ having the characteristic equations $\chi_1(x_1,\ldots,x_n)=1,\ldots,\chi_r(x_1,\ldots,x_n)=1$, then

$$(50)\qquad \Phi(x_1,\ldots,x_n) = \prod_{t=1}^{r}\chi_t(x_1,\ldots,x_n).$$

Remark 1. It is more easy to determine the characteristic function of a system of linear equations and inequalities as in § 1 c, then to compute the product of the different characteristic equations corresponding to its constraints. Therefore, if we have a system consisting of both linear and nonlinear conditions, we compute the characteristic functions of the nonlinear conditions separately, the characteristic functions of the subsystem of linear conditions, and finally their product.

Remark 2. Let us consider a system Σ whose characteristic equation is Φ. If, after obtaining Φ, we are ulteriorly given a further system Σ' which is to be fulfilled, and if we denote by Φ' its characteristic function, then the characteristic function of the completed system $\{\Sigma,\Sigma'\}$ is simply $\Phi.\Phi'$.

Example 8. Let us solve the system consisting of the nonlinear inequalities (47) from Example 7 and of the linear "sub-system" (15) from Example 3. The characteristic function of (47) is the function Φ_8 in formula (49), while the characteristic function of (15) is the function:

(16) $$\Phi_3 = \bar{x}_1 x_2 x_3 \bar{x}_4 x_5 \bar{x}_6 \bar{x}_8 \cup \bar{x}_1 x_2 x_3 \bar{x}_4 x_5 x_7 \bar{x}_8 \cup x_1 x_2 \bar{x}_3 \bar{x}_4 \bar{x}_5 \bar{x}_6 \bar{x}_8 .$$

Hence, in view of Corollary 1, the characteristic function of the augmented system in (47) & (15) is

(51) $$\Phi_9 = \Phi_3 \Phi_8 = \bar{x}_1 x_2 x_3 \bar{x}_4 x_5 \bar{x}_6 \bar{x}_8 \cup \bar{x}_1 x_2 x_3 \bar{x}_4 x_5 x_7 \bar{x}_8 .$$

§ 4. The Characteristic Function for Logical Conditions

In several practical problems we are faced with mathematical programs containing logical conditions imposed on the variables (see, for instance, G.B.DANTZIG /3/, F.RADÓ /19/ , L.NÉMETI /15/, L.NÉMETI and F.RADÓ /16/).

In this section we shall briefly examine systems of pseudo-Boolean equations and/or inequalities containing logical conditions.

For this sake, let us consider two pseudo-Boolean systems, $\Sigma'(x_1,\dots,x_n)$ and $\Sigma''(x_1,\dots,x_n)$ whose characteristic functions are $\Phi_{\Sigma'}(x_1,\dots,x_n)$ and $\Phi_{\Sigma''}(x_1,\dots,x_n)$.

If $\Sigma' \& \Sigma''$ denote the problem of finding the values of $(x_1,\dots,x_n)$ which satisfies both Σ' and Σ'', then Theorem 2 states that the characteristic function $\Phi_{\Sigma' \& \Sigma''}$ of $\Sigma' \& \Sigma''$ is

(52) $$\Phi_{\Sigma' \& \Sigma''} = \Phi_{\Sigma'} \Phi_{\Sigma''}$$

Similar results are obviously valid for other logical problems. For instance:

1. Disjunction of Σ' and Σ'', briefly $\Sigma' \vee \Sigma''$: finding the values of $(x_1, \ldots, x_n)$ which fulfil at least one of the systems Σ', Σ''. The characteristic function $\Phi_{\Sigma' \vee \Sigma''}$ of $\Sigma' \vee \Sigma''$ (i.e., the Boolean function which has the value 1 if and only if $x_1, \ldots, x_n$ fulfil $\Sigma' \vee \Sigma''$) is

(53) $$\Phi_{\Sigma' \vee \Sigma''} = \Phi_{\Sigma'} \cup \Phi_{\Sigma''} .$$

An analogous result holds for the logical disjunction of more then two systems.

2. Negation of Σ', briefly $\neg \Sigma'$: finding the values of $(x_1, \ldots, x_n)$ which do not satisfy Σ'. The characteristic function of $\neg \Sigma'$ is

(54) $$\Phi_{\neg \Sigma'} = \overline{\Phi}_{\Sigma'} .$$

3. Difference of Σ' and Σ'', briefly $\Sigma' \& \neg \Sigma''$: finding the values of $(x_1, \ldots, x_n)$ which satisfy Σ' but not Σ''. The characteristic function of $\Sigma' \& \neg \Sigma''$ is

(55) $$\Phi_{\Sigma' \& \neg \Sigma''} = \Phi_{\Sigma'} \overline{\Phi}_{\Sigma''} .$$

4. Symetric Difference of Σ' and Σ'', briefly $\Sigma' \nabla \Sigma''$: finding the values of $(x_1, \ldots, x_n)$ which fulfil ons of the systems Σ', Σ'', but not both. The characteristic function of $\Sigma' \nabla \Sigma''$ is

(56) $$\Phi_{\Sigma' \nabla \Sigma''} = \Phi_{\Sigma'} \overline{\Phi}_{\Sigma''} \cup \overline{\Phi}_{\Sigma'} \Phi_{\Sigma''} .$$

5. <u>Conditioning of $\sum''$ by $\sum'$</u>, briefly $\sum' \longrightarrow \sum''$: finding those values of $(x_1,\dots,x_n)$ which either do not satisfy $\sum'$, or satisfy both $\sum'$ and $\sum''$. The characteristic function of $\sum' \longrightarrow \sum''$ is

(57) $$\Phi_{\Sigma' \to \Sigma''} = \overline{\Phi}_{\Sigma'} \cup \Phi_{\Sigma''} .$$

Similar results can be obtained immediately for other logical conditions ("neither-nor", "If and only if", etc.).

Example 9. If $\sum'$ stands for the single inequality

(47.2) $$3x_1 - 2x_2\bar{x}_6 + 14x_5\bar{x}_6\bar{x}_8 + 2x_1x_2x_3 - 7x_8 \geqslant -8$$

and $\sum''$ denotes the single inequality

(47.4) $$2x_3 + 3x_5 - \bar{x}_5\bar{x}_6 + 4x_6\bar{x}_7x_8 - 2x_5x_6x_7x_8 \geqslant 1,$$

then the corresponding characteristic functions, determined in Example 7, are

(48'.2) $$\varphi_2 = x_1 \cup \bar{x}_2 \cup x_6 \cup \bar{x}_8$$

and

(48.4) $$\varphi_4 = x_3 \cup x_5 \cup x_6\bar{x}_7x_8,$$

respectively.

Then, taking into account Definition 1 and the above results, we see that the vector $(x_1,\dots,x_8)$:

1) satisfies at least one of the inequalities (47.2) and (47.4) if and only if

(58) $$\varphi_2 \cup \varphi_4 = x_1 \cup \bar{x}_2 \cup x_3 \cup x_5 \cup x_6 = 1 ;$$

2) does not satisfy (47.2) if and only if

(59) $$\overline{\varphi}_2 = \bar{x}_1x_2\bar{x}_6x_8 = 1 ;$$

3) satisfies (47.2), but not (47.4), if and only if

(60) $$\varphi_2\overline{\varphi}_4 = \bar{x}_3\bar{x}_5(x_1 \cup \bar{x}_2 \cup x_6 \cup \bar{x}_8)(\bar{x}_6 \cup x_7 \cup \bar{x}_8) = 1\ ;$$

4) satisfies one of the inequalities (47.2)and (47.4) but not both, if and only if

(61) $$\varphi_2\overline{\varphi}_4 \cup \overline{\varphi}_2\varphi_4 = \bar{x}_3\bar{x}_5(x_1 \cup \bar{x}_2 \cup x_6 \cup \bar{x}_8)(\bar{x}_6 \cup x_7 \cup \bar{x}_8) \cup \cup\ \bar{x}_1 x_2\bar{x}_6 x_8(x_3 \cup x_5) = 1;$$

5) either does not satisfy (47.2) or satisfies both (47.2) and (47.4), if and only if

(62) $$\overline{\varphi}_2 \cup \varphi_4 = \bar{x}_1 x_2 \bar{x}_6 x_8 \cup x_3 \cup x_5 \cup x_6 \bar{x}_7 x_8 = 1.$$

§ 5. Solving the Characteristic Equation

In the previous section we have seen that the problem of solving a system of pseudo-Boolean equations and inequalities may be reduced to that of solving its characteristic equation, which is a Boolean equation.

There are numerous methods for solving Boolean equations*). Here we shall present a procedure which offers the possibility of directly obtaining all the solutions grouped into families of solutions.

The method consists simply in writting the equation

*) See, for instance G.BIRKHOFF /1/, M.CARVALLO /2/,Ju.I.GRIGORIAN /5/, M.ITOH /6/, J.KLÍR /10/, R.S. LEDLEY /11/, L. LÖWENHEIM /12/, K.K.MAITRA /13/, M.NADLER and B.ELSPAS/14/, J.POSTLEY /17/, N.ROUCHE /20/, S.RUDEANU /21/, /22/, /23/, E.L.SCHUBERT /24/, W.SEMON /25/, A.SVOBODA and K.ČULIK/26/, A.ŽELEZNIKAR /27/, H.ZEMANEK /28/, or any standard book on Boolean algebra.

in the form

(63) $$C_1 \cup \ldots \cup C_p = 1,$$

where the C_h 's are elementary conjunctions i.e.

(64) $$C_h = x_{h_1}^{\alpha_{h_1}} \ldots x_{h_{m(h)}}^{\alpha_{h_{m(h)}}} \quad (h=1,\ldots,p);$$

these conjunctions define p families of solutions

(65) $$\mathcal{F}_h : x_{h_1} = \alpha_{h_1}, \ldots, x_{h_{m(h)}} = \alpha_{h_{m(h)}}, x_{h_{m(h)+1}}, \ldots$$
$$\ldots, x_{h_n} \text{ arbitrary } (h=1,\ldots,p),$$

which cover all the solutions of (63).

Indeed, it is obvious that each $\mathcal{F}_h$ is a family of solutions; conversely, each solution $(x_1^*,\ldots,x_n^*)$ belongs to at least one of the families $\mathcal{F}_1,\ldots,\mathcal{F}_p$, because otherwise we would have for $(x_1^*,\ldots,x_n^*)$:

$$C_1 = \ldots = C_p = 0,$$

contradicting thus (63).

Example 10. It was shown in Example 5 that the characteristic equation of the pseudo-Boolean inequality

(28) $$7x_1x_2x_3 + 5x_2x_4x_6x_7x_8 - 4x_3x_8 - 2\bar{x}_1x_4x_8 - x_4\bar{x}_5x_6 \leq 3$$

is

$$\Phi_5 = \bar{x}_1 \cup \bar{x}_2 \cup \bar{x}_3\bar{x}_4 \cup \bar{x}_3\bar{x}_6 \cup \bar{x}_3\bar{x}_7 \cup \bar{x}_3\bar{x}_8 \cup \bar{x}_4\bar{x}_8 \cup$$
$$\cup \bar{x}_6x_8 \cup \bar{x}_7x_8 = 1.$$

Hence its families of solutions are

- Table 9 -

No.	x_1	x_2	x_3	x_4	x_5	x_6	x_7	x_8
1	0	-	-	-	-	-	-	-
2	-	0	-	-	-	-	-	-
3	-	-	0	0	-	-	-	-
4	-	-	0	-	-	0	-	-
5	-	-	0	-	-	-	0	-
6	-	-	0	-	-	-	-	0
7	-	-	-	0	-	-	-	1
8	-	-	-	-	-	0	-	1
9	-	-	-	-	-	-	0	1

The same procedure offers immediately the solutions of all the problems discussed in the previous examples.

§ 6. Irredundant Solutions of the Characteristic Equation

In the previous section a method was given yielding all the solutions of the characteristic equation, grouped into families. However, the method does not assure the irredundancy of the obtained list, i.e. those solutions which belong to different families appear several times in our list. For instance, the solution $x_1=x_2=x_3=x_4=x_6=x_7=x_8=0$, $x_5=1$ in Table 9 belongs to the families 1,2,3,4,5,6 and hence, developping Table 9 into the explicit list of all the solutions, the above solution will appear 6 times.

Therefore it may be desired to have a procedure for

transforming the original families in such a way as to obtain a system of families which:

1) contain all the solutions;

2) are pairwise disjoint, i.e. the same solution cannot belong to more than one family.

The technique we shall indicate in order to sdve this problem is based on the following

Remark 3. If

(13) $$\Phi = C_1 \cup \ldots \cup C_p$$

is a disjunctive form of the characteristic function corresponding to the families $\mathcal{F}_1, \ldots, \mathcal{F}_p$ (see Theorem 1), then the above property 2 is equivalent to

(66) $$C_i C_j = 0 \text{ for } i \neq j.$$

Hence the above problem may be re-formulated as follows: if the original form (13) of the Boolean function Φ does not satisfy (66), find an equivalent disjunctive form

(67) $$\Phi = D_1 \cup \ldots \cup D_q$$

of Φ so that

(68) $$D_h D_k = 0 \text{ for } h \neq k;$$

then (67) may be called the "disjointed form" of Φ.

We start the discussion with the "linear" case :

(69) $$\Phi = x_1^{\alpha_1} \cup x_2^{\alpha_2} \cup x_3^{\alpha_3} \cup \ldots \cup x_p^{\alpha_p} .$$

LEMMA 1. <u>The disjointed form of (69) is</u>

(70) $$\Phi = x_1^{\alpha_1} \cup x_1^{\bar{\alpha}_1} x_2^{\alpha_2} \cup x_1^{\bar{\alpha}_1} x_2^{\bar{\alpha}_2} x_3^{\alpha_3} \cup \ldots \cup x_1^{\bar{\alpha}_1} \ldots x_{p-1}^{\bar{\alpha}_{p-1}} x_p^{\alpha_p} .$$

LEMMA 2. <u>If</u>

$$\Phi_h = C_{h1} \cup C_{h2} \cup \ldots \cup C_{hm(h)} \quad (h = 1,\ldots,r) \tag{71.h}$$

<u>are disjointed forms, then</u>

$$\Phi = \prod_{h=1}^{r} \Phi_h = \bigcup_{i_p,\ldots,i_r} C_{1i_1} C_{2i_2} \cdots C_{ri_r} \tag{72}$$

<u>is also a disjointed form.</u>

LEMMA 3. <u>The procedure indicated in Part I, § 4 for obtaining the families of solutions of a linear system leads to a disjointed form of the corresponding characteristic function.</u>

Now, a disjointed form of the characteristic function $\varphi(x_1,\ldots,x_n)$ of a single pseudo-Boolean equation or inequality may be obtained as follows:

We find, as in § 1 the characteristic function $\psi(y_1,\ldots,y_m)$ of the associated linear equation (inequality).

Each y_i is a product of variables x_j with or without bars, while $\bar{y}_i$ is a disjunction of variables x_j without or with bars. We replace the y_i and $\bar{y}_i$ by their expressions, we apply Lemma 1 to each disjunction corresponding to an $\bar{y}_i$ appearing in $\psi(y_1,\ldots,y_m)$, after which we perform all the multiplications.

Further, if we want to obtain a disjointed form of the characteristic function Φ of a system, then we simply multiply the disjointed forms of the characteristic functions φ of the different equations and inequalities of the system.

It is not difficult to prove the following:

THEOREM 3. <u>The above procedure leads to a disjointed form of the characteristic function $\Phi(x_1,\dots,x_n)$ of the given system.</u>

Example 12. Let us consider the system

(73.1) $$2x_1x_2x_4 - 4x_5x_6 + 3x_3 \leqslant 2.$$

(73.2) $$4x_1x_3x_5 + 6x_2x_4x_6 \geqslant 4.$$

Denoting

(74) $$x_5x_6=y_1,\ x_1x_2x_4=y_2,\ x_2x_4x_6=y_3,\ x_1x_3x_5=y_4,$$

we can writte the inequalities (73) in the form

(74.1) $$4y_1 + 3\bar{x}_3 + 2\bar{y}_2 \geqslant 3,$$

(74.2) $$6y_3 + 4y_4 \geqslant 4.$$

The characteristic functions of these inequalities are

(74.1) $$\psi_1(y_1,x_3,y_2) = y_1 \cup \bar{y}_1\bar{y}_3,$$

(74.2) $$\psi_2(y_3,\ y_4) = y_3 \cup \bar{y}_3 y_4\ ,$$

hence

(75.1) $$\varphi_1(x_1,\dots,x_6) = x_5x_6 \cup (\bar{x}_5 \cup \bar{x}_6)\bar{x}_3,$$

(75.2) $$\varphi_2(x_1,\dots,x_6) = x_2x_4x_6 \cup (\bar{x}_2 \cup \bar{x}_4 \cup \bar{x}_6)x_1x_3x_5.$$

Making the $\bar{y}_i$ disjoint and performing the multiplications we obtain the disjointed forms of φ_1 and φ_2 :

(75'.1) $$\varphi_1(x_1,\dots,x_6) = x_5x_6 \cup (\bar{x}_5 \cup x_5\bar{x}_6)\ \bar{x}_3 = x_5x_6 \cup \bar{x}_3\bar{x}_5 \cup \bar{x}_3x_5\bar{x}_6,$$

(75'.2) $$\varphi_2(x_1,\dots,x_6) = x_2x_4x_6 \cup (\bar{x}_2 \cup x_2\bar{x}_4 \cup x_2x_4\bar{x}_6)x_1x_3x_5 = x_2x_4x_6 \cup x_1\bar{x}_2x_3x_5 \cup x_1x_2x_3\bar{x}_4x_5 \cup x_1x_2x_3x_4x_5\bar{x}_6.$$

Therefore a disjointed form of the characteristic function of the system (73) is

$$(76)\quad \Phi_9(x_1,\dots,x_6) = (x_5x_6 \cup \bar{x}_3\bar{x}_5 \cup \bar{x}_3x_5\bar{x}_6)(x_2x_4x_6 \cup \cup x_1\bar{x}_2x_3x_5 \cup x_1x_2x_3\bar{x}_4x_5 \cup x_1x_2x_3x_4x_5\bar{x}_6) = = x_2x_4x_5x_6 \cup x_1\bar{x}_2x_3x_5x_6 \cup x_1x_2x_3\bar{x}_4x_5x_6 \cup x_2\bar{x}_3x_4\bar{x}_5x_6,$$

corresponding to the following complete system of disjoint families of solutions:

- Table 10 -

No.	x_1	x_2	x_3	x_4	x_5	x_6
1	-	1	-	1	1	1
2	1	0	1	-	1	1
3	1	1	1	0	1	1
4	-	1	0	1	0	1

§ 7. The Pseudo-Boolean Form of the Characteristic Function

The characteristic function of a pseudo-Boolean system is a Boolean function. However, in Part III it will be necessary to have a pseudo-Boolean expression of the characteristic function, i.e. an expression using only the arithmetical operations " + ", " - ", and possibly the negation " $^{-}$ " of single variables.

The following identities are well-known:

$$a_1 \cup a_2 = a_1 + a_2 - a_1a_2,$$

$$a_1 \cup a_2 \cup a_2 = a_1 + a_2 + a_3 - a_1a_2 - a_1a_3 - a_2a_3 + a_1a_2a_3,$$

. . . .

etc., which permit to transform every disjunctive form of a Boolean function into a pseudo-Boolean one.

If

(77) $$\Phi = C_1 \cup \ldots \cup C_m$$

is a disjointed form of the Boolean function Φ, then $C_iC_j=0$ for all $i \neq j$, and the above identities show that relation (77) may by simply written in the pseudo-Boolean form

(86') $$\Phi = C_1 + \ldots + C_m.$$

Example 13. The characteristic function Φ_9 given by formula (76) in Example 12 can be written in the pseudo-Boolean form

(78) $$\Phi_9(x_1,\ldots,x_6) = x_2x_4x_5x_6 + x_1\bar{x}_2x_3x_5x_6 + x_1x_2x_3\bar{x}_4x_5x_6 + x_2\bar{x}_3x_4x_5x_6.$$

§ 8. Computational Status

The methods developed in this Part were testet yet only by hand computation and the results seam to be extremely encouraging.

The programming of the procedure for a MECIPT computer is in progress.

References

1. G.BIRKHOFF : Lattice Theory. Amer.Math.Soc.Coll.Publ. New York 1948 (reprint 1961).
2. M.CARVALLO : Principes et applications de l'analyse booléenne. Gauthier-Villars, Paris 1965.
3. G.B.DANTZIG : Linear Programming and Extensions, Ch.26.Princeton Univ.Press, Princeton, 1963.
4. R.FORTET : L'algèbre de Boole et ses applications en recherche opérationnelle. Cahiers Centre Etudes Rech.Opér., 1, No.4, 5-36 (1959).
5. Ju.I.GRIGORIAN : Algorithm for the Solution of Logical Equations (in Russian). Jurn.vycislit.mat.i mat.fiziki, 2, 186-189 (1962).
6. M.ITOH : On Boolean Equations with Many Unknowns and the Generalized Poretsky's Formula. Rev.Univ. Nac. Tucumán,12, 1o7-112 (1959).
7. P.L.IVANESCU : Systems of Pseudo-Boolean Equations and Inequalities. Bull.Acad. Polon.Sci.Ser.Math.Astronom.Phys., 12, 673-68o (1964).
8. P.L.IVANESCU : The Method of Succesive Eliminations for Pseudo-Boolean Equations. Bull. Acad.Polon.Ser.Sci.Math. Astronom.Phys., 12, 681-683 (1964).
9. P.L.IVANESCU and S.RUDEANU : The Theory of Pseudo-Boolean Programming. I. Linear Pseudo-Boolean Equations and Inequalities.SIAM Journal (in press).
10. J.KLÍR : Solutions of System of Boolean Equations (in Czech.), Apl. Mat. 7, 265-273 (1962).
11. R.S.LEDLEY : Digital Computers and Control Engineering. Mc. Graw Hill, New York, 1960.
12. L.LÖWENHEIM : Über Auflösungsproblem im logischen Klassenkalkul. Sitzungsber. Berl.Math.Geselschaft,7,89-94(19o8).
13. K.K.MAITRA : A Map Approach to the Solution of a Class of Boolean Functional Equations. Communic. Electr. No. 59, 34-36 (1962).
14. M.NADLER and B.ELSPAS : The Solution of Simultaneous Boolean Equations. IRE Trans.Communication Theory 7,No.3 (1960).

15. L.NÉMETI : Das Reihenfolgsproblem in der Fertigungsprogrammierung und Linearplanung mit logischen Bedingungen. Mathematica (Cluj), 6(29), 87-99, (1964).

16. L.NÉMETI, F.RADÓ : Ein Wartezeitproblem in der Programmierung der Produktion. Mathematika (Cluj), 5(28), 65-95 (1963).

17. J.POSTLEY : A Method for the Evaluation of a System of Boolean Algebraic Equations. Math.Tables and Other Aids to Computations, 9, 5-8 (1955).

18. F.RADÓ : Linear Programming with Logical Conditions (in Romanian). Comunicările Acad.RPR, 13, 1o39-1o42 (1963).

19. F.RADÓ : Un algorithme pour résoudre certains problèmes de programmation mathématiques. Mathematica (Cluj),6(29), 1o5-116 (1964).

20. N.ROUCHE : Some Properties of Boolean Equations.IRE Trans. Electronic Computers, 7, 291-298 (1958).

21. S.RUDEANU : On the Solution of Boolean Equations by the Löwenheim Method (in Romanian). Stud.Cerc.Mat., 13, 295-3o8 (1962).

22. S.RUDEANU : Remarks on Motinori Goto's Papers on Boolean Equations. Rev.Roumaine Math.Pures Appl.,10, 311-317 (1965).

23. S.RUDEANU : Irredundant Solutions of Boolean and Pseudo-Boolean Equations. Rev.Roumaine Math.Pures Appl. (in press).

24. E.L.SCHUBERT : Simultaneous Logical Equations. Comm. and Electronics, No.46, 1o8o-1o83 (1960).

25. W.SEMON : A Class of Boolean Equations. Sperry Rand. Research Corp., SRCC-RR-62-17, August 1962.

26. A.SVOBODA and K.ČULIK : An Algorithm for Solving Boolean Equations (in Russian).Avtomat.i Telemeh.25, 374-381, (1964).

27. A.ŽELEZNIKAR : Behandlung logistischer Probleme mit Ziffernrechner. Glasnik Mat.-Fiz.-Astronom,17, 171-179 (1962).

28. H.ZEMANEK : Die Lösung von Gleichungen in der Schaltalgebra. Arch. Elektr. Ubertragung, 12, 35-44 (1958).

Part III

MINIMIZATION OF PSEUDO-BOOLEAN FUNCTIONS

In this part, we give an algorithm for finding the minimum of a pseudo-Boolean function as well as its minimizing points. This procedure is, in fact, a combination of the dynamic programming approach with Boolean techniques (§ 1). In §§ 2-5, the method is extended to the case when the variables have to fulfil certain pseudo-Boolean conditions (equations, inequalities,logical conditions); the method proposed in § 4 seems to be the most efficient.

We mention that the special problem of minimizing a linear pseudo-Boolean function with linear or nonlinear constraints has a surprisingly simple solution (§ 2),but the general (nonlinear) case is also solved in an efficient way.

The importance of programming with bivalent (0,1) variables was repeatedly pointed out by G.B. DANTZIG /3/, /4/, R.FORTET /5/,/6/ and by many other authors. The fact that any problem of integer(linear or nonlinear) programming may be reduced to one of bivalent programming, is well-known.

The method proposed in this paper require the solution of certain system of pseudo-Boolean equations and inequalities.

This can be done with the procedures given in Parts I and II.

A. MINIMA WITHOUT CONSTRAINTS

§ 1. The Basic Algorithm

Definition 1. A vector

(1) $$(x_1^*,\ldots,x_n^*)\in B_2^n$$

is a minimizing point of the pseudo-Boolean function $f(x_1,\ldots,x_n)$, if

(2) $$f(x_1^*,\ldots,x_n^*)\leq f(x_1,\ldots,x_n)$$

for any $(x_1,\ldots,x_n)$ in B_2^n and the real $f(x_1^*,\ldots,x_n^*)$ is the global minimum - or, simply, the minimum - of f.

We see from the definition that

(3.i) $$f(x_1^*,\ldots,x_n^*)\leq f(x_1^*,\ldots,x_{i-1}^*, \overline{x_i^*}, x_{i+1}^*,\ldots,x_n^*)$$

for all $i = 1,2,\ldots,n$.

The conditions (3.i) are hence necessary for $(x_1^*,\ldots,x_n^*)$ to be a minimizing point, but not sufficient: they characterize the local minima of the function f. However - and this is the basic idea of the algorithm we are going to describe here for obtaining the minimizing points - if we successively solve the inequalities (3.1),(3.2),...,etc., introducing in each inequality the parametric solution of the preceding ones, then we obtain all the minimizing points and only them. This method was first given by I.ROSENBERG and the present authors in /11/ , starting from an idea of R.FORTET /5/, /6/.

We describe below the proposed algorithm,followed by an example and by some computational remarks.

Description of the algorithm

Our algorithm is made up of two main stages: in the first one, the minimum of the given pseudo-Boolean function $f(x_1,..,x_n)$ is found, while in the second one all the minimizing points are determined.

In fact, the recursive procedure we are going to propose here, may be viewed as a dynamic approach to bivalent programming (as it was shown in /9/).

First stage. Let us denote, for the sake of recurrency,

$$f_1(x_1,...,x_n) = f(x_1,...,x_n). \tag{4.1}$$

It was seen in Part I that any pseudo-Boolean function is a polynomial, linear in each of the variables. Hence the function f_1 may be written in the form

$$f_1(x_1,...,x_n) = x_1 g_1(x_2,...,x_n) + h_1(x_2,...,x_n) \tag{5.1}$$

and the inequality (3.1) becomes

$$(x_1 - \bar{x}_1)\, g_1(x_2,...,x_n) \leqslant 0. \tag{6.1}$$

For satisfying (6.1) it is necessary and sufficient that x_1 should satisfy the following conditions: if $g_1(x_2,...,x_n) < 0$ then $x_1 = 1$; if $g_1(x_2,...,x_n) > 0$, then $x_1 = 0$; if $g_1(x_2,...,x_n) = 0$, then $x_1 = p_1$ = arbitrary bivalent parameter (for, in this case, the inequality is satisfied for each value of x_1).

Let us now introduce a Boolean function φ_1 depending on the variables $x_2,...,x_n$ and on a new variable p_1 :

$$\varphi_1 : B_2^n \longrightarrow B_2, \tag{7.1}$$

defined by

(8.1) $$\varphi_1(p_1,x_2,\ldots,x_n) = \begin{cases} 1, & \text{if } g_1(x_2,\ldots,x_n) < 0, \\ 0, & \text{if } g_1(x_2,\ldots,x_n) > 0, \\ p_1, & \text{if } g_1(x_2,\ldots,x_n) = 0. \end{cases}$$

In view of the above discussions, we have:

LEMMA 1. For every $(p_1,x_2,\ldots,x_n) \in B_2^n$, if

(9.1) $$x_1 = \varphi_1(p_1,x_2,\ldots,x_n),$$

then the vector $(x_1,x_2,\ldots,x_n) \in B_2^n$ satisfies (6.1) and, conversely, if $(x_1,x_2,\ldots,x_n) \in B_2^n$ is a solution of (6.1), then there exists an element $p_1 \in B_2$ so that relation (9.1) holds.

Let us now notice that the function φ_1 may be also written in the form

(10.1) $$\varphi_1(p_1,x_1,\ldots,x_n) = \varphi_1'(x_2,\ldots,x_n) \cup p_1\varphi_1''(x_2,\ldots,x_n),$$

where the functions φ_1' and φ_1'' are defined as follows:

(11.1) $$\varphi_1'(x_2,\ldots,x_n) = \begin{cases} 1, & \text{if } g_1(x_2,\ldots,x_n) < 0, \\ 0, & \text{if } g_1(x_2,\ldots,x_n) \geq 0, \end{cases}$$

and

(12.1) $$\varphi_1''(x_2,\ldots,x_n) = \begin{cases} 1, & \text{if } g_1(x_2,\ldots,x_n) = 0, \\ 0, & \text{if } g_1(x_2,\ldots,x_n) \neq 0. \end{cases}$$

We see that φ_1' and φ_1'' are the characteristic functions (as defined in Part II, § 1) of the inequality $g_1 < 0$ and of the equation $g_1 = 0$, respectively. We have given in Part II, § 2, Example 6) a method (for the usual case when the coefficients are integers) for directly obtaining these two functions from the characteristic function of the inequality $g_1 \leq 0$. Hence a Boolean expression of φ_1 may be constructed and this will be used in the second stage of the algorithm.

Let us now construct the function

$$(42)\qquad f_2(x_2,\dots,x_n) = f_1(\varphi_1'(x_2,\dots,x_n),\ x_2,\dots,x_n).$$

To do this, we express the function φ_1' in pseudo-Boolean form (i.e., using the arithmetical operations + , - , and possibly the negation of single variables), as it was done in § 6 of Part II, and introduce it into (5.1) :

$$f_2(x_2,\dots,x_n) = \varphi_1'(x_2,\dots,x_n)g_1(x_2,\dots,x_n)+h_1(x_2,\dots,x_n).$$

We proceed now with the function f_2 in the same way as with f_1, etc. At the step i of the first stage we are faced with a pseudo-Boolean function f_i of n-i+1 variables $x_i, x_{i+1}, \dots, x_n$:

$$(1.i)\qquad f_i : B_2^{n-i+1} \longrightarrow R,$$

which is written in the form

$$(5.i)\qquad f_i(x_i,x_{i+1},\dots,x_n) = x_i g_i(x_{i+1},\dots,x_n) + h_i(x_{i+1},\dots,x_n),$$

and we have to solve the inequality

$$(6.i)\qquad (x_i - \bar{x}_i)g_i(x_{i+1},\dots,x_n) \leq 0.$$

To do this, we introduce a Boolean function φ_i of n-i+1 variables $p_i, x_{i+1},\dots,x_n$:

$$(7.i)\qquad \varphi_i : B_2^{n-i+1} \longrightarrow B_2,$$

defined by

$$(8.i)\qquad \varphi_i(p_i,x_{i+1},\dots,x_n) = \begin{cases} 1, & \text{if } g_1(x_{i+1},\dots,x_n) < 0, \\ 0, & \text{if } g_i(x_{i+1},\dots,x_n) > 0, \\ p_i, & \text{if } g_i(x_{i+1},\dots,x_n) = 0, \end{cases}$$

or, equivalently, by

(10.i) $$\varphi_i(p_i, x_{i+1}, \dots, x_n) = = \varphi_i'(x_{i+1}, \dots, x_n) \cup p_i \varphi_i''(x_{i+1}, \dots, x_n)$$

where

(11.i) $$\varphi_i'(x_{i+1}, \dots, x_n) = \begin{cases} 1, & \text{if } g_i(x_{i+1}, \dots, x_n) < 0, \\ 0, & \text{if } g_i(x_{i+1}, \dots, x_n) \geqslant 0 \end{cases}$$

and

(12.i) $$\varphi_i''(x_{i+1}, \dots, x_n) = \begin{cases} 1, & \text{if } g_i(x_{i+1}, \dots, x_n) = 0, \\ 0, & \text{if } g_i(x_{i+1}, \dots, x_n) \neq 0. \end{cases}$$

We write down the formula

(9.i) $$x_i = \varphi_i(p_i, x_{i+1}, \dots, x_n)$$

and pass to the (i+1)-th step, constructing the function

$$f_{i+1}(x_{i+1}, \dots, x_n) = = f_i(\varphi_i'(x_{i+1}, \dots, x_n), x_{i+1}, \dots, x_n).$$

Continuing in this way, we obtain at step n a function $f_n(x_n)$ which is written in the form

(5.n) $$f_n(x_n) = x_n g_n + h_n$$

where g_n and h_n are reals.

We have to solve the inequality

(6.n) $$(x_n - \bar{x}_n) g_n \leqslant 0;$$

using the function

(8.n) $$\varphi_n(p_n) = \begin{cases} 1, & \text{if } g_n < 0, \\ 0, & \text{if } g_n > 0, \\ p_n, & \text{if } g_n = 0, \end{cases}$$

the solution is

(9.n) $$x_n = \varphi_n(p_n).$$

It can be proved that the minimum of the original function f is

(4.n+1) $$f_{min} = f_{n+1} = f_n(\varphi'_n),$$

where the constant φ'_n is given by

(11.n) $$\varphi'_n = \begin{cases} 1, \text{ if } g_n < 0, \\ 0, \text{ if } g_n \geqslant 0. \end{cases}$$

The first stage of the algorithm has come to an end.

Second stage. Let us consider the formula

(9.n) $$x_n = \varphi_n(p_n)$$

and introduce it into

(9.n-1) $$x_{n-1} = \varphi_{n-1}(p_{n-1}, x_n),$$

obtaining thus

(9'.n-1) $$x_{n-1} = \varphi_{n-1}(p_{n-1}, \varphi_n(p_n));$$

then we introduce the expressions (9.n) and (9'.n-1) into

(9.n-2) $$x_{n-2} = \varphi_{n-2}(p_{n-2}, x_{n-1}, x_n)$$

obtaining the new relation

(9'.n-2) $$x_{n-2} = \varphi_{n-2}(p_{n-2}, \varphi_{n-1}(p_{n-1}, \varphi_n(p_n)), \varphi_n(p_n)),$$

etc.

We obtain in this way a sequence of relations of the form

(13.n) $$x_n = \psi_n(p_n),$$

(13.n-1) $$x_{n-1} = \psi_{n-1}(p_{n-1}, p_n).$$

$$\cdots\cdots\cdots\cdots$$

(13.1) $$x_1 = \psi_1(p_1, \ldots, p_{n-1}, p_n).$$

The following remark is very important for the effectiveness of the algorithm.

Remark 1. Practice has shown that the functions φ_i do actually depend only on a small number of parameters $p_{i_1},\dots,p_{i_m}$.

It will be proved below (Theorem 1) that, giving to the parameters $p_{i_1},\dots,p_{i_m}$ which actually appear in formula (13), all possible values, we obtain all the minimizing points of the function f and only them.

Example 1. Let us minimize the pseudo-Boolean function

(14.1) $f=f_1= 2x_1+ 3x_2- 7x_3- 5x_1x_2x_3+ 3x_2x_4+ 9x_4x_5- 2x_1x_5,$

which may be also written in the form

$$f_1 = x_1(2-5x_2x_3 - 2x_5) + 3x_2 - 7x_3 + 3x_2x_4 + 9x_4x_5.$$

Hence

(15.1) $$g_1 = 2 - 5x_2x_3 - 2x_5$$

so that using the methods given in Part II, we obtain

(16.1) $$\varphi_1' = x_2x_3, \quad \varphi_1'' = (\bar{x}_2 \cup \bar{x}_3)x_5,$$

therefore formula (9.1) becomes

(17.1) $$x_1 = \varphi_1' \cup p_1\varphi_1'' = x_2x_3.$$

We replace x_1 by $\varphi_1' = x_2x_3$ in f_1 and obtain the function

(14.2) $f_2 = - 3x_2x_3 + 3x_2 - 7x_3 + 3x_2x_4 + 9x_4x_5 - 2x_2x_3x_5.$

Hence

(15.2) $$g_2 = 3 - 3x_3 + 3x_4 - 2x_3x_5$$

so that

(16.2) $$\varphi_2' = x_3\bar{x}_4x_5, \quad \varphi_2'' = x_3\bar{x}_4\bar{x}_5;$$

therefore relation (9.2) becomes

(17.2) $$x_2 = \varphi_2' \cup p_2 \varphi_2'' = x_3\bar{x}_4x_5 \cup p_2x_3\bar{x}_4\bar{x}_5.$$

We replace x_2 by $\varphi_2' = x_3\bar{x}_4x_5$ in f_2 and obtain the function

(14.3) $$f_3 = -7x_3 + 9x_4x_5 - 2x_3\bar{x}_4x_5.$$

Hence

(15.3) $$g_3 = -7 - 2\bar{x}_4 ,$$

so that

(16.3) $$\varphi_3' = 1, \quad \varphi_3'' = 0,$$

and relation (9.3) becomes

(17.3) $$x_3 = \varphi_3' \cup p_3 \varphi_3'' = 1.$$

We replace x_3 by $\varphi_3 = 1$ in f_3 and get the function

(14.4) $$f_4 = -7 + 9x_4x_5 - 2\bar{x}_4x_5 = -7 - 2x_5 + 11x_4x_5,$$

for $\bar{x}_4 = 1 - x_4$.

Hence

(15.4) $$g_4 = 11x_5,$$

(16.4) $$\varphi_4' = 0, \quad \varphi_4'' = \bar{x}_5 ,$$

and relation (9.4) is reduced to

(17.4) $$x_4 = \varphi_4' \cup p_4 \varphi_4'' = p_4\bar{x}_5.$$

We replace x_4 by $\varphi_4' = 0$ in f_4 and obtain

(14.5) $$f_5 = -7 - 2x_5,$$

hence

(15.5) $$g_5 = -2,$$

(16.5) $$\varphi_5' = 1, \quad \varphi_5'' = 0$$

and relation (9.5) becomes

(17.5) $$x_5 = \varphi_5' \cup p_5 \varphi_5'' = 1.$$

The minimum value of the function f is obtained by putting $x_5 = \varphi_5' = 1$ in f_5:

(14.6) $$f_{min} = f_6 = -9.$$

It remains to determine the minimizing points.Relation (17.5) is simply

(18.5) $$x_5 = 1 ;$$

introducing this value into (17.4), we get

(18.4) $$x_4 = 0 ;$$

further, relation (17.3) is simply

(18.3) $$x_3 = 1 ;$$

introducing the above values into (17.2), we have

(18.2) $$x_2 = 1 ;$$

and finally, from (17.1) we get

(18.1) $$x_1 = 1.$$

Therefore the single minimizing point is (1,1,1,0,1).

Example 2. The minimization of the pseudo-Boolean function

(19) $$2x_1 + 3x_2 - 7x_3 - 5x_1x_2x_3 + 3x_2x_4 + 9x_4x_5$$

may be carried out as in Table 1.

- Table 1 -

No.	f_i	g_i	φ_i'	φ_i''	x_i
1	$2x_1+3x_2-7x_3-5x_1x_2x_3+$ $+3x_2x_4+9x_4x_5$	$2-5x_2x_3$	x_2x_3	0	x_2x_3
2	$-3x_2x_3+3x_2-7x_3+$ $+3x_2x_4+9x_4x_5$	$3-3x_3+3x_4$	0	$x_3\bar{x}_4$	$p_2x_3\bar{x}_4$
3	$-7x_3+9x_4x_5$	-7	1	0	1
4	$-7+9x_4x_5$	$9x_5$	0	$\bar{x}_5$	$p_4\bar{x}_5$
5	-7	0	0	1	p_5
6	$f_6 = f_{min} = -7$				

Hence $f_{min} = -7$ and the minimizing points are

$$x_5 = p_5,$$
$$x_4 = p_4\bar{p}_5,$$
$$x_3 = 1,$$
$$x_2 = p_2(\bar{p}_4 \cup p_5),$$
$$x_1 = p_2(\bar{p}_4 \cup p_5).$$

For the various values of the parameters p_2, p_4, p_5, we obtain <u>all</u> the minimizing points $(x_1,\dots,x_5)$ of f_1 :

(20.1) $x_1 = 0,\ x_2 = 0,\ x_3 = 1,\ x_4 = 0,\ x_5 = 0;$

(20.2) $x_1 = 0,\ x_2 = 0,\ x_3 = 1,\ x_4 = 0,\ x_5 = 1;$

(20.3) $x_1 = 0,\ x_2 = 0,\ x_3 = 1,\ x_4 = 1,\ x_5 = 0;$

(20.4) $x_1 = 1,\ x_2 = 1,\ x_4 = 1,\ x_4 = 0,\ x_5 = 0;$

(20.5) $x_1 = 1,\ x_2 = 1,\ x_3 = 1,\ x_4 = 0,\ x_5 = 1.$

We are now going to give below a series of remarks

concerning the above described algorithm. Remarks 2, 3, 9 and 10 are aimed to accelerate the computations.

Remark 2. In the preceding discussion, we have started by elimitating the variable x_1, followed by x_2, x_3 etc. In fact, there is no special reason for preferring this order, and we may adopt any other one which seems to be more appropriate. For instance, if we denote by n_i the number of terms in g_i, where $f = x_i g_i + h_i$, then it seems convenient to start with the elimination of that variable x_{i_0} for which $n_{i_0} = \min_i n_i$.

Remark 3. If, after eliminating the variables x_{i_1}, $x_{i_2}, \ldots, x_{i_m}$, where $m < n$, we obtain f_{m+1} = constant, then the variables $x_{i_{m+1}}, \ldots, x_{i_n}$ are arbitrary (because, for any $q=1,2,\ldots,n-m$ we have $g_{i_{m+q}} = 0$, hence $\varphi'_{i_{m+q}} = 0$, $\varphi''_{i_{m+q}} = 1$; therefore $x_{i_{m+q}} = \varphi'_{i_{m+q}} + p_{i_{m+q}} \varphi''_{i_{m+q}} = p_{i_{m+q}}$).

Remark 4. The function φ_i may be also written in the form

$$(21.i) \quad \varphi_i(p_i, x_{i+1}, \ldots, x_n) = \varphi'_i(x_{i+1}, \ldots, x_n) \cup p_i \varphi'''_i(x_{i+1}, \ldots, x_n)$$

where φ'''_i is the function defined by

$$(22.i) \quad \varphi'''_i(x_{i+1}, \ldots, x_n) = \begin{cases} 1, & \text{if } g_i(x_{i+1}, \ldots, x_n) \leq 0, \\ 0, & \text{if } g_i(x_{i+1}, \ldots, x_n) > 0 \end{cases}$$

(for it is obvious that $\varphi'''_i = \varphi'_i \cup \varphi''_i$, hence (10.i) implies

$$\varphi_i = \varphi'_i \cup p_i \varphi''_i = \varphi'_i \cup p_i \varphi'_i \cup p_i \varphi''_i = \varphi'_i \cup p_i \varphi'''_i).$$

Remark 5. We have $\varphi'_i(x_{i+1}, \ldots, x_n) = \varphi_i(0, x_{i+1}, \ldots, x_n)$, so that relation (4.i+1) becomes

$$(23.i+1) \quad f_{i+1}(x_{i+1}, \ldots, x_n) = f_i(\varphi_i(0, x_{i+1}, \ldots, x_n), x_{i+1}, \ldots, x_n).$$

Remark 6. The function f_{i+1} may be also written in the form

$$(24.i+1)\quad f_{i+1}(x_{i+1},\dots,x_n)=f_i(\varphi_i(1,x_{i+1},\dots,x_n),x_{i+1},\dots,x_n),$$

for relations (8.i) and (11.i) imply that the function

$$\varphi_i g_i = \varphi_i' g_i = \begin{cases} g_i, & \text{if } g_i<0, \\ 0, & \text{if } g_i\geqslant 0, \end{cases}$$

does not depend on p_i, hence formula $f_i = x_i g_i + h_i$ implies, in turn

$$f_{i+1} = f_i(\varphi_i', x_{i+1},\dots,x_n) = \varphi_i' g_i + h_i = \varphi_i g_i + h_i =$$
$$= \varphi_i(0,x_{i+1},\dots,x_n)g_i(x_{i+1},\dots,x_n) + h_i(x_{i+1},\dots,x_n)=$$
$$= \varphi_i(1,x_{i+1},\dots,x_n)g_i(x_{i+1},\dots,x_n) + h_i(x_{i+1},\dots,x_n)=$$
$$= f_i(\varphi_i(1,x_{i+1},\dots,x_n),x_{i+1},\dots,x_n).$$

Remark 7. The above Remarks 5 and 6 show that the function f_{i+1} may be written in the form

$$(25.i+1)\quad f_{i+1}(x_{i+1},\dots,x_n)=f_i(\varphi_i(p_i,x_{i+1},\dots,x_n),x_{i+1},\dots,x_n),$$

whatever the element p_i may be (0 or 1).

Remark 8. The function f_{i+1} may be also written in the form

$$(26.i+1)\quad f_{i+1}(x_{i+1},\dots,x_n)=f_i(\varphi_i'''(x_{i+1},\dots,x_n),x_{i+1},\dots,x_n)$$

(for, as it was shown in Remark 4, $\varphi_i''' = \varphi_i \cup \varphi_i'' = \varphi_i(1,x_{i+1},\dots$ $\dots,x_n)$ so that relation (24.i+1) may be written in the form (26.i+1)).

Remark 9. When the problem is to find only the minimum of the function (and not the minimizing points), then the algorithm reduces to its first stage, which, in its turn, becomes more simple. Namely, at the i-th stage it suffices to determine either φ_i' or φ_i''' and to obtain f_{i+1} using either formula (4.i+1) or formula (26.i+1) respectively.

Remark 10. When the problem is to find not all, but a single minimizing point, then the first stage of the algorithm is performed as in Remark 9, while the second stage is also simplified in an obvious way, the role of the functions φ_i being played by the simpler functions φ_i' or φ_i'''.

It is easy to prove:

LEMMA 2. The vector $(x_1^*,\dots,x_n^*)$ is a minimizing point of the function $f_1(x_1,x_2,\dots,x_n)$ if and only if the following two conditions hold:

(α) there exists an element $p_1^* \in B_2$ such that $x_1^* = \varphi_1(p_1^*,x_2,\dots,x_n^*)$;

(β) the vector $(x_2,\dots,x_n)$ is a minimizing point of the function $f_2(x_2,\dots,x_n)$.

Lemma 2 is basic in proving

THEOREM 1. The vector $(x_1^*,\dots,x_n^*)$ is a minimizing point of the pseudo-Boolean function $f(x_1,\dots,x_n)$, if and only if there exist*) values $p_1^*,\dots,p_n^*$ in B_2, so that

(27.1) $$x_1^* = \varphi_1(p_1^*,x_2^*,\dots,x_n^*),$$

(27.2) $$x_2^* = \varphi_2(p_2^*,x_3^*,\dots,x_n^*),$$

$$\dots\dots\dots\dots$$

(27.n) $$x_n^* = \varphi_n(p_n^*).$$

The minimum of $f(x_1,\dots,x_n)$ is f_{n+1} defined as in (4.n+1).

---

*) Notice that Theorem 1 does not contradict Remark 1. For, if the function φ_i does not actually depend on p_i, then, of course, there exists a value p_i^* so that relation (31.i) take place.

B.MINIMA WITH CONSTRAINTS

In this division we shall examine several alternative methods for solving the problem of the determination af all the minimizing points of a pseudo-Boolean function, whose variables are subject to pseudo-Boolean constraints.

The main tools will be the solution of pseudo-Boolean equations, inequalities and logical conditions,as it was done in Parts I and II, and the algorithm for minimizing pseudo-Boolean functions, described in § 1.

The next two sections deal with special problems. In § 2 we deal with the case when the conditions are of the form $f_j = 0$, where each f_j is either non-negative,or non-positive. In § 3 we study the problem of minimizing linear functions, with linear or nonlinear constraints. The method in that section permits a surprisingly quick detection of the minimizing points.

The general problem of the determination of the minimizing points of an arbitrary pseudo-Boolean function with arbitrary constraints, will be dealt with in § 3, 4 and 5 in different alternative ways. The best of these methods seems to be that given in § 4.

§ 2. Lagrangeian Multipliers

Let us consider the problem of minimizing the pseudo-Boolean function with <u>integer</u> values

$$f(x_1,\dots,x_n) \tag{28}$$

subject to the pseudo-Boolean equations

$$(29) \qquad f_j(x_j,\dots,x_n) = 0 \qquad (j=1,\dots,m),$$

where each function f_j is <u>integer</u> - valued and either non-negative, or non-positive. Without loss of generality, we may suppose that all $f_j \geqslant 0$, i.e.

$$(30.j) \qquad f_j(x_1,\dots,x_n) \geqslant 0 \text{ for all } (x_1,\dots,x_n) \in B_2^n \ (j=1,\dots,m).$$

Problems of this type arise frequently in applications; many a problem of the theory of graphs /8/,/10/, and of networks /7/, or of switching theory /12/, may be formulated in this form.

The method we describe below was first given in /12/.

Let us now consider an expression of the function f (as a polynomial in $x_1,\dots,x_n$ and possibly in $\bar{x}_1,\dots,\bar{x}_n$ and let S^+ and S^- be the sums of the positive and of the negative coefficients of f, respectively.

We have

THEOREM 2. (α) <u>If the vector</u> $X^* = (x_1^*,\dots,x_n^*)$ <u>minimizes f, being subject to the constraints (33), then X* minimizes also the pseudo-Boolean function</u>

$$(31) \quad F(x_1,\dots,x_n) = f(x_1,\dots,x_n) + (S^+ - S^- + 1)\sum_{j=1}^{m} f_j(x_1,\dots,x_n).$$

(β) <u>If X* minimizes F and</u>

$$(32) \qquad F(x_1^*,\dots,x_n^*) \leqslant S^+,$$

<u>then X* minimizes f under the conditions (33)</u>.

(γ) <u>If X* minimizes F and</u>

$$(33) \qquad F(x_1^*,\dots,x_n^*) > S^+,$$

<u>then the conditions (29) are inconsistent</u>.

Remark 11. The above theorem remains valid if we replace S^+ and S^- by any upper bound and lower bound of f, respectively.

Corollary 1. For finding all the points in B_2^n which minimize the function f under the conditions $f_j = 0$, we apply the algorithm in § 1 to the function F.

If the value $F_{n+1} = F_{min} > S^+$, then the problem has no solutions. If $F_{n+1} = F_{min} \leq S^+$, then this value is the minimum of the function f under the restrictions $f_j = 0$, and the solutions of the considered problem coincide with the minimizing points of F (which may be obtained by applying the second stage of the algorithm).

Example 3. Let us minimize the function

$$f = 2x_1x_3 - x_2\bar{x}_3 + 4x_3x_4x_5 - x_1x_5 \tag{34}$$

with the condition

$$x_1x_2 + 2x_2x_3 + \bar{x}_4x_5 = 0. \tag{35}$$

Here $S^+ = 6$, $S^- = -2$, hence we have to minimize the function

$$F = 2x_1x_3 - x_2\bar{x}_3 + 4x_3x_4x_5 - x_1x_5 + 9(x_1x_2 + 2x_2x_3 + \bar{x}_4x_5). \tag{36}$$

We find, with the basic algorithm, that $F_{min} = F_6 = -1 < S^+$, hence the problem is consistent. The minimizing points are

$$x_1 = 0, \quad x_2 = 1, \quad x_3 = 0, \quad x_4 = p_5 \cup p_4, \quad x_5 = p_5. \tag{37}$$

Remark 12. In many a problem the conditions are expressed by Boolean equations

$$\varphi_j = c_{j_1} \cup \ldots \cup c_{j_{k(j)}} = 0, \quad (j=1,\ldots,m) \tag{38.j}$$

where $c_{j_1}, c_{j_2}, \ldots, c_{j_{k(j)}}$ are elementary conjunctions. Since the equations (38) may be also written in the pseudo-Boolean form

$$(39.j) \qquad f_j = c_{j_1} + c_{j_2} + \ldots + c_{j_{k(j)}} = 0 \quad (j=1,\ldots,m),$$

the functions b_j being non-negative, we can apply the above described method.

Example 4. /10/. The maximization of the pseudo-Boolean function

$$(40) \qquad f = x_1 + x_2 + x_3 + x_4 + x_5 + x_6$$

under the condition

$$(41) \qquad x_1x_4 \cup x_1x_5 \cup x_1x_6 \cup x_2x_4 \cup x_3x_6 \cup x_5x_6 = 0$$

is equivalent to the minimization of the function $-f$ under the restriction (41), or to the minimization of the unrestricted pseudo-Boolean function

$$(42) \qquad F = - x_1-x_2-x_3-x_4-x_5-x_6 + 7(x_1x_4 + x_1x_5 + x_1x_6 + x_2x_4 + x_3x_6 + x_5x_6).$$

Applying the basic algorithm, we find $F_{min} = F_7 = -3$ (hence the sought maximum is $+3$) and the minimizing points

$$(43) \qquad x_1=\bar{p}_5, \quad x_2=\bar{p}_4 \cup \bar{p}_5, \quad x_3=1, \quad x_4=p_4p_5, \quad x_5=p_5, \quad x_6=0.$$

Remark 13. Consider the case when only part of the constraints satisfy the conditions (30). Reasoning as in Theorem 2, we may transform the problem into another one, having those constraints which fulfil (30) introduced into the function to be minimized, and maintaining the other constraints in their original form.

§ 3. Minimization of Linear Pseudo-Boolean Functions

The minimization of a linear pseudo-Boolean function

(44) $$c_1x_1 + \dots + c_nx_n$$

may be performed without any difficulty; indeed, its minimizing points are defined by

(45) $$x_i^* = \begin{cases} 1, \text{ if } & c_i < 0; \\ 0, \text{ if } & c_i > 0; \\ p_i, \text{if } & c_i = 0; \end{cases}$$

where p_i is an arbitrary parameter in B_2.

Example 5. It is obvious that the minimizing points of

(46) $$2 + 3x_1 - 2x_2 - 5x_3 + 2x_6 - x_7$$

are

(47) $$x_1=0, \quad x_2=1, \quad x_3=1, \quad x_4=p_4, \quad x_5=p_5, \quad x_6=0, \quad x_7=1$$

where p_4 and p_5 are arbitrary parameters in B_2. Hence, the minimum of (46) is - 6.

The minimization of a linear pseudo-Boolean function

(44) $$f(x_1,\dots,x_n) = f(X) = c_1x_1 + \dots + c_nx_n$$

subject to certain (linear, or nonlinear) constraints*) may be performed in a similar way.

Namely, the method will comprise three steps:

1. Determination of the solutions of the system of restrictions, grouped into families of solutions (as in Parts I and II) $\mathcal{F}_1,\dots,\mathcal{F}_p$.

*) As it was shown by R.FORTET /5/,/6/,any pseudo-Boolean program may be reduced to a program of this type, by introduction of certain supplementary variables.

2. For each family $\mathcal{F}_k$ of solutions, determination of

(45) $$\min_{X\in\mathcal{F}_k} f(X)$$

and of those points $X^o \in \mathcal{F}_k$ for which

(46) $$f(X^o) = \min_{X\in\mathcal{F}_k} f(X)$$

3. Determination (by direct checking) of

(47) $$\min_{k=1,\dots,p} \; \min_{X\in\mathcal{F}_k} f(X)$$

and of those points X^* in B_2^n for which

(48) $$f(X^*) = \min_{k=1,\dots,p} \; \min_{X\in\mathcal{F}_k} f(X)$$

It remains now to indicate hos to perform step 2.

The vectors $X = (x_1,\dots,x_n)$ belonging to a family $\mathcal{F}_k$ of solutions are characterized by the fact that the values x_i are fixed for those i which are contrained in a certain set I_k of indices:

(49) $$i \in I_k \text{ implies } x_i = x_i^* = \text{fixed} \quad (0 \text{ or } 1).$$

while x_i remain arbitrary for $i \notin I_k$.

Reasoning as above, it is easy to see that the points X^o satisfying (46) are given by the following formula:

(50) $$x_i = \begin{cases} x_i^*, & \text{if } i \in I_k, \\ 1, & \text{if } i \notin I_k \text{ and } c_i < 0, \\ 0, & \text{if } i \notin I_k \text{ and } c_i > 0, \\ p_i, & \text{if } i \notin I_k \text{ and } c_i = 0. \end{cases}$$

where p_i are arbitrary parameters in B_2.

Example 6. Let us minimize

(51) $$2 + 3x_1 - 2x_2 - 5x_3 + 2x_4 + 4x_6$$

with the constraints

(52.1) $$2x_1 - 3x_2 + 5x_3 - 4x_4 + 2x_5 - x_6 \leqslant 2$$

(52.2) $$4x_1 + 2x_2 + x_3 + 8x_4 - x_5 - 3x_6 \geqslant 4.$$

The families of solutions of (52), determined as in Part I, § 4, are:

- Table 2 -

No.	x_1	x_2	x_3	x_4	x_5	x_6
1	-	-	0	1	-	-
2	-	1	1	1	-	-
3	0	0	1	1	0	-
4	0	0	1	1	1	1
5	1	0	1	1	0	1
6	1	1	0	0	-	0
7	1	0	0	0	0	0

where the dashes indicate the arbitrary variables.

Putting in Table 2 instead of dashes the values given by (50) we obtain

- Table 3 -

No.	x_1	x_2	x_3	x_4	x_5	x_6	Value of (51)
1	0	1	0	1	p_1	0	2
2	0	1	1	1	p_2	0	-3
3	0	0	1	1	0	0	-1
4	0	0	1	1	1	1	3
5	1	0	1	1	0	1	6
6	1	1	0	0	p_6	0	3
7	1	0	0	0	0	0	5

Hence the sought minimum is -3 and it is attained in the points (0,1,1,1,0,0) and (0,1,1,1,1,0).

Example 7. Let us minimize

(53) $$2 + 3x_1 - 2x_2 - 5x_3 + 2x_4 + 4x_6$$

with the nonlinear constraints

(54.1) $$x_1x_2 + 4\bar{x}_1x_3 - 3x_2x_3x_5 + 6\bar{x}_2x_4x_6 \geqslant -1$$

(54.2) $$3x_2x_4 - 5\bar{x}_1\bar{x}_3\bar{x}_5 + 4x_4x_6 \geqslant 1.$$

The families of solutions of the constraints, obtained as in Part II, are:

- Table 4 -

No.	x_1	x_2	x_3	x_4	x_5	x_6
1	0	1	1	1	-	0
2	0	-	1	1	-	1
3	0	1	0	1	1	0
4	0	-	0	1	1	1
5	0	1	0	1	0	1
6	1	0	-	1	-	1
7	1	1	0	1	-	-
8	1	1	1	1	0	-

Putting in Table 5 instead of dashes the values indicated by (50) we find:

- Table 5 -

No.	x_1	x_2	x_3	x_4	x_5	x_6	Value of (53)
1	0	1	1	1	p_1	0	-3
2	0	1	1	1	p_2	1	+1
3	0	1	0	1	1	0	+2
4	0	1	0	1	1	1	+6
5	0	1	0	1	0	1	+6
6	1	0	1	1	p_6	1	+6
7	1	1	0	1	p_7	0	+5
8	1	1	1	1	0	0	o

Hence the sought minimum is -3 and it is attained in the points (0,1,1,1,0,0) and (0,1,1,1,1,0).

Accelerated Method for Linear Pseudo-Boolean Programming.

The above described procedure comprises three steps : determination of all the solutions to the constraints , determination of the partially minimizing points corresponding to the various families of solutions, and choice of the minimizing points among the partially minimizing ones. This technique takes no advantage, in the first (and most cumbersome) step , of the informations supplied by the objective function.

In order to utilize more completely the data of the problem, we can proced as follows.

We add a supplementary constraint

$$f\ (x_1,\ldots,x_n) \leqslant M_r,$$

where f is the objective function, while M_r is a coefficient to be defined below.

At the beginning of the process (r=0), M_o is either equal to the value $f(x_1^{(o)},\ldots,x_n^{(o)})$ in a point $(x_1^{(o)},\ldots,x_n^{(o)})$ satisfying the constraints - in case such a point is a priori known - or equal to an upper bound of the function f (for instance, the sum of its positive coefficients).

Let $G^{(1)}$ be the first family of solutions to the augmented system of constraints, and let $(x_1^{(1)},x_2^{(1)},\ldots,x_n^{(1)})$ be a partially minimizing point of the function f corresponding to $G^{(1)}$. We put $M_1 = f(x_1^{(1)},x_2^{(1)},\ldots,x_n^{(1)})$ and continue the bifurcation process with the system consisting of the original constraints and of the new inequality $f \leqslant M_1$. Etc.

Obviously, the last coefficient M_s is the sought minimum, while the minimizing points are all those partially minimizing points $(x_1^{(r)},\ldots,x_n^{(r)})$ for which $M_r = M_s$.

Thus the above modified algorithm avoids the determination of all families of solutions to the constraints, substantially reducing in this way the amount of necessary computations.

§ 4. Minimization Using Families of Solutions

The method described in the previous section may be extended to the case of arbitrary (linear or nonlinear) objective functions. Namely, the knowledge of the p families of solutions to the constraints allows to transform the original problem into (at most !) p minimization problems for unrestricted pseudo-Boolean functions, each of which has less variables than the original function*).

This procedure seems to be the best of the different approaches proposed in this paper for solving the general problem.

Let us consider again the problem of minimizing a pseudo-Boolean function

*) M.CARVALLO /2/ proposes to test all the solutions of the constraints and to choose those for which the minimum is reached.

(55) $$f(x_1,\dots,x_n)$$

subject to the constraints

(56) $$f_j(x_1,\dots,x_n) \lesseqqgtr 0 \quad (j=1,\dots,m)$$

and let $\mathcal{F}_1,\dots,\mathcal{F}_p$ be the families of solutions of (56).

We may proced now as follows.

(1.k.) Introduce the fixed variables of the family $\mathcal{F}_k$ into (56);

(2.k.) Minimize the unrestricted pseudo-Boolean function obtained at (1.k.); let v^k be the corresponding minimum.

(3). Choose those points for which $v^{k_o} = \min_{k=1,\dots,p} v^k$.

Example 8. The minimization of the pseudo-Boolean function

(57) $$3x_1\bar{x}_2 - 8\bar{x}_1x_3x_6 + 4x_2x_5\bar{x}_6 + 7\bar{x}_5x_6 + 3x_4 - 5x_4x_5x_6$$

under the constraints

(52.1) $$2x_1 - 3x_2 + 5x_3 - 4x_4 + 2x_5 - x_6 \leq 2,$$

(52.2) $$4x_1 + 2x_2 + x_3 + 8x_4 - x_5 - 3x_6 \geq 4,$$

leads to the minimization of 7 unrestricted pseudo-Boolean functions corresponding to the 7 families of solutions.

Proceeding as indicated above, we obtain Table 6, showing that the minimum is -12 and it is reached in the points (0,1,1,1,0,1) and (0,0,1,1,0,1).

Example 9. The minimization of the same pseudo-Boolean function (57) under the constraints

(54.1) $$x_1x_2 + 4\bar{x}_1x_3 - 3x_2x_3x_5 + 6\bar{x}_2x_4x_6 \geq -1,$$

(54.2) $$3x_2x_4 - 5\bar{x}_1\bar{x}_3\bar{x}_5 + 4x_4x_6 \geq 1.$$

- Table 6 -

No.	Families of Solutions $x_1x_2x_3x_4x_5x_6$	Function to be minimized	Partially minimizing points x_1 x_2 x_3 x_4 x_5 x_6	Partial minimum
1	- - 0 1 - -	$3x_1\bar{x}_2-8\bar{x}_1x_3x_6+$ $+4x_2x_5x_6-7\bar{x}_5x_6+$ $+3x_4-5x_4x_5x_6$	p_1p_2 p_2 0 1 0 1	-4
2	- 1 1 1 - -	$-8\bar{x}_1x_6+4x_5\bar{x}_6-$ $-7\bar{x}_5x_6+3-5x_5x_6$	0 1 1 1 0 1	-12
3	0 0 1 1 0 -	$3-15x_6$	0 0 1 1 0 1	-12
4	0 0 1 1 1 1	---	0 0 1 1 1 1	-10
5	1 0 1 1 0 1	---	1 0 1 1 0 1	-1
6	1 1 0 0 - 0	$4x_5$	1 1 0 0 0 0	0
7	1 0 0 0 0 0	---	1 0 0 0 0 0	3

leads to the minimization of 8 unrestricted pseudo-Boolean functions corresponding to the 8 families of solutions.

Proceeding as indicated above, we obtain Table 7, showing that the minimum is -12 and it is reached in the same points as in Example 10 : (0,0,1,1,0,1) and (0,1,1,1,0,1).

- Table 7 -

No.	Families of solutions $x_1x_2x_3x_4x_5x_6$	Functions to be minimized	Partially minimizing points x_1 x_2 x_3 x_4 x_5 x_6	Partial minimum
1	0 1 1 1 - 0	$4x_5 + 3$	0 1 1 1 0 0	3
2	0 - 1 1 - 1	$-10 - 2\bar{x}_5$	0 p_2^2 1 1 0 1	-12
3	0 1 0 1 1 0	---	0 1 0 1 1 0	7
4	0 - 0 1 1 1	---	0 p_2^4 0 1 1 1	-2
5	0 1 0 1 0 1	---	0 1 0 1 0 1	-7
6	1 0 - 1 - 1	$1 - \bar{x}_5$	1 0 p_3^6 1 0 1	0
7	1 1 0 1 - -	$4x_5\bar{x}_6-7\bar{x}_5x_6+$ $+ 3 - 5x_5x_6$		
8	1 1 1 1 0 -	$- 7x_6 + 3$	1 1 1 1 0 1	-4

§ 5. Other Methods for Solving Pseudo-Boolean Programs

α). R.FORTET /5/ has proposed the following approach to the general problem of minimizing a pseudo-Boolean function under constraints: find parametric arithmetical expressions ("codage") of the solutions to the constraints, introduce them into the objective function and minimize the pseudo-Boolean function with independent variables obtained in this way. In the same paper, the parametrization of certain important types of constraints is given.

The concrete method we suggest for achieving this purpose, consiste of the following four steps:

1. Determine the characteristic equation of the constraints, as in Parts I and II.

2. Find a parametric solution of the characteristic equation, using one of the well known methods for solving Boolean equations.

3. Write the solution obtained at the step 2 in a pseudo-Boolean form.

4. Introduce this solution into the objective function and minimize the new function obtained in this way, by means of the basic algorithm.

In view of the results obtained in Parts I and II and in § 1 of this paper, the above algorithm may be applied to any bivalent program*).

β). The basic algorithm given in § 1 of this Part for the minimization of a pseudo-Boolean function without restrictions, may be extended to the case when the variables are subject to certain pseudo-Boolean conditions.

This modified recursive procedure reduces at each stage the number of variables with one, assuring the fulfilment of the constraints. Its basic idea goes back to dynamic programming.

γ). The method of Lagrangeian multipliers, given in § 2 for <u>integer</u> - valued pseudo-Boolean functions, may be extended to the general case when the constraints need not satisfy conditions (30). For, in this case, let $\Phi(x_1,\ldots,x_n)$ be the characteristic function (written in a pseudo-Boolean form) of the system of constraints. As Φ takes only the values 0 and 1, it ful-

*) An interesting method using the solution of Boolean equations was proposed by M.CARVALLO /2/, using an idea of P.CAMION /1/.

fils condition (30). Therefore the problem of minimizing f under the given constraints, reduces to that of minimizing the restriction-free pseudo-Boolean function $f + \lambda\varphi$, for a sufficiently large λ.

References

1. P.CAMION : Une méthode de résolution par l'algèbre de Boole des problèmes combinatoires où interviennent des entiers. Cahiers du Centre d'Etudes de Recherches Opérationnelle, 2, 234-289 (1960).
2. M.CARVALLO: Principes et applications de l'analyse booléenne. Gauthiers-Villars, Paris, 1965.
3. G.B.DANTZIG : On the Significance of Solving Linear Programming Problems with Some Integer Variables. Econometrica, 28, No.1, 30-44 (1960).
4. G.B.DANTZIG : Linear Programming and Extensions. Ch.26,Princeton University Press, Princeton, 1963.
5. R.FORTET : L'algèbre de Boole et ses applications en recherche opérationnelle. Cahiers du Centre d'Etudes de Recherche Opérationnelle, 1, No.4, 5-36 (1959).
6. R.FORTET : Application de l'algèbre de Boole en recherche opérationnelle. Revue Française de Recherche Opérationnelle, No.14, 1960.
7. P.L.IVANESCU : Some Network Flow Problems Solved by Pseudo-Boolean Programming. Operations Research, 13, No.3, 388-399 (1965).

8. P.L.IVANESCU : Pseudo-Boolean Programming with Special Restraints. Applications to Graph Theory. Elektronische Informationsverarbeitung und Kybernetik (EIK) 3, 167-185, (1965).

9. P.L.IVANESCU : Dynamic Programming with Bivalent Variables. Lecture at the Symposium on Applications of Mathematics to Economics, Smolenice (Czechoslovakia) June 1965 (to appear in Publ.Inst.Math. Belgrade).

10. P.L.IVANESCU and I.ROSENBERG : Application of Pseudo-Boolean Programming to the Theory of Graphs. Z.Wahrscheinlichkeitstheorie, 3, No.2, 167-176 (1964).

11. P.L.IVANESCU, I.ROSENBERG and S.RUDEANU : On the Determination of the Minima of Pseudo-Boolean Functions (in Romanian). Studii şi Cercetări Mat., 14, No.3, 359 - 364 (1963).

12. P.L.IVANESCU, I.ROSENBERG and S.RUDEANU : An Application of Discrete Linear Programming to the Minimization of Boolean Functions (in Russian). Revue Math.Pures et Appl., 8, No.3, 459-475 (1963).

13. P.L.IVANESCU and S.RUDEANU : The Theory of Pseudo-Boolean Programming. I. Linear Pseudo-Boolean Equations and Inequalities. SIAM Journal (in press).

14. P.L.IVANESCU and S.RUDEANU : The Theory of Pseudo-Boolean Programming.II. Nonlinear Pseudo-Boolean Equations and Inequalities. SIAM Journal (in press).

PART IV

FRACTIONAL BIVALENT PROGRAMMING

The problem of minimizing the cost of production of a certain item, i.e. the quotient of the (usually linear) function representing the total cost, by the (usually linear) function representing the produced amount of that item, is an outstanding example of what is termed "hyperbolic" or "fractional" programming. Methods for solving this problem were given by B.MARTOS /3/, A.CHARNES and W.W.COOPER /1/, and W. DINKELBACH /2/.

We are going now to present algorithms for solving problems of this type for the case of bivalent variables. In other words, the problem is to minimize a function of the form

$$F = \frac{a_o + a_1x_1 + \dots + a_nx_n}{b_o + b_1x_1 + \dots + b_nx_n}, \tag{1}$$

where the variables x_h $(h = 1,\dots,n)$ may take only the values 0 and 1.

We shall examine here only the case - frequently appearing in practice - when

(2') $$b_o > 0,$$

(2") $$b_i \geq 0 \quad (i=1,\ldots,n),$$

(as a matter of fact, it can be easily shown that this restriction is not essential).

Let I and J be the sets of all indices $i > 0$ and $j > 0$ satisfying

(3) $$\frac{a_i}{b_i} \leq \frac{a_o}{b_o}$$

and

(4) $$\frac{a_o}{b_o} < \frac{a_j}{b_j},$$

respectively.

The following algorithm is leading to all the minimizing points of F:

<u>Algorithm I.</u>

1. Whenever $a_h = 0$, $b_h > 0$, put $x_h = 1$.
2. Whenever $a_h > 0$, $b_h = 0$, put $x_h = 0$.
3. Determine the sets I and J.
4. For each $j \in J$, put $x_j = 0$.
5. Determine the first index i_1 for which $\frac{a_{i_1}}{b_{i_1}} = \min_{i \in I} \frac{a_i}{b_i}$.

Case (α). If $\frac{a_{i_1}}{b_{i_1}} < \frac{a_o}{b_o}$, put $x_{i_1} = 1$, transform a_o into $a_o + a_{i_1}$, transform b_o into $b_o + b_{i_1}$ and perform again the steps 3, 4 and 5. Case (β). If $\frac{a_{i_1}}{b_{i_1}} = \frac{a_o}{b_o}$, then for each $i \in I$ put $x_i = p_i$, where p_i is an arbitrary parameter.

Example 1. Minimize

$$(5)\quad F = \frac{3+2x_1+4x_2+x_3+2x_4+9x_5+6x_6+12x_7+8x_8+2x_9+3x_{10}+3x_{11}+x_{12}}{6+x_1+8x_2+3x_3+5x_4+15x_5+10x_6+25x_7+18x_8+6x_9+3x_{10}+7x_{11}}$$

1. There is no h with $s_h = 0$, $b_h > 0$.

2. Since $a_{12} = 1$, $b_{12} = 0$, we take $x_{12} = 0$; there is no other h with $a_h > 0$, $b_h = 0$.

3. We have to determine the sets J and I for the

$$F_1 = \frac{3+2x_1+4x_2+x_3+2x_4+9x_5+6x_6+12x_7+8x_8+2x_9+3x_{10}+3x_{11}}{6+x_1+8x_2+3x_3+5x_4+15x_5+10x_6+25x_7+18x_8+6x_9+3x_{10}+7x_{11}}$$

We have $\frac{a_o}{b_o} = \frac{1}{2}$ and

$J = \{1,5,6,10\}$,
$I = \{2,3,4,7,8,9\}$.

4. We put $x_1 = x_5 = x_6 = x_{10} = 0$.

5. The first index i_1 for which $\frac{a_{i_1}}{b_{i_1}} = \min_{i\in I} \frac{a_i}{b_i}$ is $i_1 = 3$. Since $\frac{a_3}{b_3} = \frac{1}{3} < \frac{a_o}{b_o}$, we are in the case ($\alpha$) and so we take $x_3 = 1$. Now we are faced with the new function

$$F_2 = \frac{4 + 4x_2 + 2x_4 + 12x_7 + 8x_8 + 2x_9 + 3x_{11}}{9 + 8x_2 + 5x_4 + 25x_7 + 18x_8 + 6x_9 + 7x_{11}}$$

3'. We have now $\frac{a'_o}{b'_o} = \frac{4}{9}$

$J' = \{2,7\}$,
$I' = \{4,8,9,11\}$.

4'. We put $x_2 = x_7 = 0$.

5'. The single index i_1 for which $\frac{a_{i_1}}{b_{i_1}} = \min_{i\in I'} \frac{a_i}{b_i}$ is $i_1 = 9$.

We have $\frac{a_9}{b_9} = \frac{1}{3} < \frac{a'_o}{b'_o}$, so that we are again in the case (α) and $x_9 = 1$. The new function is

$$F_3 = \frac{6 + 2x_4 + 8x_8 + 3x_{11}}{15 + 5x_4 + 18x_8 + 7x_{11}} .$$

3". Now $\frac{a''_o}{b''_o} = \frac{2}{5}$ and

$J'' = \{8,11\}$,

$I'' = \{4\}$.

4". We put $x_8 = x_{11} = 0$.

5". Since $\frac{a_4}{b_4} = \min_{i \in I''} \frac{a_i}{b_i} = \frac{2}{5} = \frac{a''_o}{b''_o}$, we are in the case (β) so that x_4 is an arbitrary parameter.

The sought minimum is

$$(6) \qquad F_4 = f(0,0,1,x_4,0,0,0,0,1,0,0,0) = \frac{6 + 2x_4}{15 + 5x_4} = \frac{2}{5}$$

It can be proved that in case of a pseudo-Boolean function $F(x_1,\ldots,x_n)$ of the form (1), satisfying (2), the conditions

$$(7) \qquad T_i^* \leq 0 \quad (i=1,\ldots,n)$$

where

$$(8) \qquad T_i^* = (x_i^* - \overline{x_i^*})(b_i \sum_{j=1}^{n} a_j x_j^* - a_i \sum_{j=1}^{n} b_j x_j^*) \quad (i=1,\ldots,n),$$

are necessary and sufficient for $(x_1^*,\ldots,x_n^*)$ to be a minimizing point of the function F. Hence we are led to the following

Algorithm II. Start with any initial vector $(x_1^+,\ldots,x_n^+)$ (the initial vector

$$(9) \qquad x_i^+ = \begin{cases} 1, \text{ if } b_i > 0 \text{ and } \frac{a_i}{b_i} < \frac{a_o}{b_o}, \\ 0, \text{ otherwise,} \end{cases}$$

seems to give a good approximation of the minimum). Compute, for this vector, the quantities T_i^+. If all $T_i^+ \leq 0$, then we have obtained the sought minimizing point. If there are positive T_i^{+}'s, then choose the greatest one, say $T_{i_o}^+$, and change $x_{i_o}^+$ into $\overline{x_{i_o}^+}$, leaving $x_1^+, \ldots, x_{i_o-1}^+, x_{i_o+1}^+, \ldots, x_n^+$ unchanged. Compute again the new T_i^{++}'s and check whether they are all non-positive or not, etc. This procedure is continued, until all T_i's become non-positive, showing that we have obtained a minimizing point.

Example 2. Applying the above algorithm to the function (5) studied in Example 1, we obtain the following table:

- Table 1 -

No.	x_1	x_2	x_3	x_4	x_5	x_6	x_7	x_8	x_9	x_{10}	x_{11}	x_{12}	T_1	T_2	T_3	T_4	T_5	T_6	T_7	T_8	T_9	T_{10}	T_{11}	T_{12}	i_o
1	0	0	1	1	0	0	1	1	1	0	1	0	-	-	-	-	-	-	+	+	-	-	+	-	7
2	0	0	1	1	0	0	0	1	1	0	1	0	-	-	-	-	-	-	-	+	-	-	+	-	8
3	0	0	1	1	0	0	0	0	1	0	1	0	-	-	-	-	-	-	-	-	-	-	+	-	11
4	0	0	1	1	0	0	0	0	1	0	0	0	-	-	-	0	-	-	-	-	-	-	-	-	

In case of restrictions, we apply the method given in parts I and II for finding their families of solutions, and determine the partially minimizing points, by means of one of the above algorithm.

Remark. Algorithm I gives all the minimizing points of (1), while algorithm II offers a single one.

References

/1/ A.CHARNES and W.W.COOPER, Programming with Linear Fractional Functionals. Naval Res. Log. Quarterly, 9 , No.3-4, 181-186 (1962).

/2/ W.DINKELBACH, Die Maximierung eines Quotienten zweier linearen Funktionen unter linearen Nebenbedingungen. Z.Wahrscheinlichkeitstheorie, 1, 141-145 (1962).

/3/ B.MARTOS, Hyperbolic Programming (in Hungarian).Mag.Tud. Akad. Mat. Kut. Intéz. Közlemenyei, 5, series B.No.4, 383-406 (1960).